Ramanujan & Ancient Indian Mathematicians

Swapan Banerjee

First Published : 2019
Reprint : 2020
Copyright © Swapan Banerjee 2019

Title ID 7841830

ISBN-10   1981156283

ISBN-13   978-1981156283

Dedicated to
Indian Philosophy and way of life
Of which
My mother Suniti Banerjee was
An integral part

Contents

Foreword	7
Acknowledgements	13
Preface	15

1. Life of Ramanujan   17 -106

1.1	Childhood of Ramanujan	22
1.2	College Student Ramanujan	27
1.3	Unemployed Ramanujan	31
1.4	Self-Reliant Ramanujan	38
1.5	The Research student Ramanujan	50
1.6	Ramanujan in England	59
1.7	S. Ramanujan, B.A.	65
1.8	Ramanujan became F.R.S.	71
1.9	Ramanujan's Return to India	82
1.10	Ramanujan's relatives and mentors after his death	93
1.11	Ramanujan's Mathematics	102
1.12	Ramanujan's Notebooks	106

2. Numerals of Different Countries   107 -152

2.1	Ancient Roman Numeration	112
2.2	Ancient Greek Numeration	114
2.3	Slavic Numeration	118
2.4	Ancient Armenian and Georgian numeration	118
2.5	Babylonian Numeration	119
2.6	Ancient Mayan Numeration	120
2.7	Ancient Chinese Numeration	121

	2.8	Ancient Egyptian Numeration	127
	2.9	Indian Numeration	129
	2.10	History of Zero	134
	2.11	Propagation of ancient Indian Mathematics	144
	2.12	Evolution of Indo-Arabic numerals	152

3. Ancient Indian Mathematicians   153- 189

	3.1	Baudhayana	155
	3.2	Manava	157
	3.3	Apastamba	158
	3.4	Panini	158
	3.5	Katyayana	158
	3.6	Aryabhata I	159
	3.7	Varahamihira	166
	3.8	Brahmagupta	167
	3.9	Bhaskara I	170
	3.10	Sridharacharya	172
	3.11	Bateswar	172
	3.12	Aryabhata II	174
	3.13	Bhaskara II	174
	3.14	Madhava	176
	3.15	Nilakantha Somayaji	177
	3.16	Radhanath Sikdar	187
	3.17	Ashutosh Mukherjee	188

4.	**Questions of Quiz game**	190
5.	**Answers of the Quiz**	192
6.	**References**	194
6.	**Appendix**	196

Foreword

If we go through our Vedas, we may say that 'Numerals' are essentially of Indian origin. It is known to us that Medhātithi (who belonged to the gotra or family of Kaṇva) extended the numerals to billions. His name is associated in Ṛk-Veda, Atharva-Veda, Yajur-Veda. He contributes to the powers of 'Ten'. It is true that we find the numerals in the Ṛk-Veda, Gopatha Brāhmaṇa, Yajur-Veda, Atharva-Veda, Taittirīya Saṁhita and other books.

In the Vājasaneyi Saṁhita verse no. (mantra) 17.2, we find the following numbers : eka ($10°$), daśa(10^1), śata (10^2), sahasra (10^3), ajutam(10^4), nijutam or laksa(10^5), prayutam (10^6), arbuda or koti (10^7), nyarbudam or abja (10^8), kharva (10^9), nikharva(10^{10}), mahāpadma(10^{11}), śaṅku(10^{12}), samudra(10^{13}), madhya(10^{14}), antya(10^{15}), parārdha(10^{16}). But, kharva (10^9), nikharva(10^{10}), mahāpadma(10^{11}) and śaṅku (10^{12}) do not occur in the original verses of Vājasaneyi Saṁhita. Mahidhara, the commentator of Vājasaneyi Saṁhita includes these four terms. In the epic Ramayana (vi: 2.28) we find the following numbers (Rama's army), $10^{10} + 10^{14} + 10^{20} + 10^{24} + 10^{30} + 10^{34} + 10^{40} + 10^{44} + 10^{52} + 10^{57} + 10^{62} + 5$.

In Gaṇita Sāra Saṁgraha, we find twenty four notational places and the last is Mahaksobha =10^{23}. In the Jyotiṣkaraṇdaka, we find the name of 250th place which is called Śīrsa-prahelika (someone says it 194th place).

The science of geometry was originated in India in connection with the construction of the altars of the Vedic sacrifices. The Śulba Sūtras are manuals for the constructions of altars. They are sections of the Kalpa Sūtras or particularly the Śrauta Sūtras. It is one of the six Vedāṅgas. The term Śulba or Śulva is derived from the root sulb or śulv, which means 'to measure' and in etymological sense it connotes 'measuring' an act of measurement. Sometimes, the term śulba means a rope or a cord.

At present only eight Śulba Sūtras are available. These are:

(1) Baudhāyana Śulba Sūtra which belongs to Kṛṣṇa Yajur-Veda. It is divided into three chapters of which 1st and 2nd chapter contain 196 Sūtras and the commentator is Dvārakānatha Yajva. The name of the book (commentary) is Śulba Dipika. The third chapter contains 323 Sūtras; the commentator is Venkateśvara Dīkṣita. The name of the book (commentary) is Śulba Mīmāṁsā.

(2) Āpastamba Śulba Sūtra belongs to Kṛṣṇa Yajur-Veda. It is divided into twenty one khandas. The total number of Sūtras are 223. The commentators are (1) Kapardisvāmī (Śulba Vyākhyā), (2) Karavindasvamī (Śulba Pradīpikā).

(3) Kātyāyana Śulba Sūtra is divided into two parts. Part 1 contains 90 Sūtras and the commentator is Rāma or Rāmachandra (Śulba Sūtra Vṛtti). Part 2 contains 46 or 48 verses. The commentator is Mahidhara (Śulba Sūtra Vivarana). Kātyāyana Śulba Sūtra belongs to Śukla Yajur-Veda.

(4) Manava Śulba Sūtra belongs to Kṛṣṇa Yajur-Veda. It is divided into four khaṇḍas.

(5) Maitrāyana Śulba Sūtra belongs to Kṛṣṇa Yajur-Veda. It is divided into four khaṇḍas.

(6) Vārāha Śulba Sūtra belongs to Kṛṣṇa Yajur-Veda. It is divided into three parts and many sections.

(7) Vādhula Śulba Sūtra belongs to Kṛṣṇa Yajur-Veda.

(8) Hiraṇya Keśī.

Baudhāyana and Āpastamba are not the discoverer of the principles of geometry. For that reason they used the terms (i) iti abhyupadiśanti i.e. it is so recognised by the authorities, (ii) iti vijñāyate i.e. thus they teach, (iii) iti uktam i.e. it has been so said.

Sometimes we see especially in Kātyāyana Śulba Sūtra the sentence "Rajju Samasaṁ" i.e. we will expand the manipulation with cords i.e. the term Rajju is given instead of Śulba.

Influence to the Brahamanic literature for geometry primarily comes from Ṛk-Veda where we find the term 'trisad hasthe' which means three places (of the Agni). These three types are Gārhapatya (circular), Āhavanīya (square), and Dakṣiṇāgni (semi-circle). Later we find the following:

(I) Ālaja-citi, (ii) Droṇa citi, (trough), (iii) Kaṅka citi, (iv) Kūrma citi (tortoise), (v) Pra-u-ga citi (isosceles triangle),

(vi) Rathacakra citi (chariot wheel), (vii) Samūhya citi, (viii) Śmaśāna citi, (ix) Ubhayataḥ pra-u-ga citi, (x) Vakrapakṣvyasta puccha Śyena citi.

In order to construct these altars the following geometrical operations are needed.
   (i)   To construct a square equal to a simple equal multiple (or submultiple) of another square.
   (ii) To construct a square equal to the sum or difference of two unequal squares.
   (iii) To transform a rectangle into a square and vice-versa.
   (iv) To construct a triangle or rhombus equal to a square etc.

Baudhāyana is the oldest Śulbakāra. His Śulba Sūtra is completed in three chapters. He gave a general enunciation of Pythagoras theorem, an approximate value of $\sqrt{2}$ correct to five places of decimal.

There are reasons to believe that Āryabhaṭa lived at Kusumpura and wrote his famous book Āryabhaṭiya there. In support of this opinion we may mention the verse no.1 of chapter II of Āryabhaṭiya. He wrote :

ब्रह्मकुशशिबुधभृगुरविकुजगुरुकोराभगराान्नमस्कृत्य ।
त्र्यार्यभटस्विह निगदति कुसुमपुरेऽभ्यर्चिते ज्ञाने ।।

Foreword

The Āryabhaṭīya deals with both mathematics and astronomy. It contains 121 stanzas in all. It is divided into four chapters i.e. four pādas.

Pāda 1 (Gītikā pāda) consists of thirteen stanzas. In this pāda Āryabhaṭa discussed bhagaṇas or number of cycles of planets, some important definitions, observed positions of nodes and uccas, elements of eccentrics and epicycles and numerical values of twenty four traditional sines in a quadrant.

Pāda 2 (Ganita pāda) consists of 33 stanzas and deals with mathematics. It contains the geometrical figures, their properties and mensuration, series, interest, simple, simultaneous, quadratic and linear indeterminate equations, square root, cube root etc.

Pāda 3 (Kālakriyā pāda) containing 25 stanzas deals with various units, determinations of the true position of moon, sun and other planets by means of eccentric cycles and epicycles etc.

Pāda 4 (Gola pāda) consists of 50 stanzas deals with the motion of the sun, moon and other planets on the celestial sphere etc.

The first section is devoted to the life and work of Ramanujan briefly. In this section a fascinating account of

Ramanujan's life is dealt with, which reads like a sad romantic novel.

In the second section, the author discusses the numerals of different countries.

The third section deals with the life and works of a few ancient and medieval Indian mathematicians.

Swapan Banerjee has discussed everything in detail in this book.

Dr Pradip Kumar Majumdar
Former Professor of Indian Astronomy,
Rabindra Bharati University, Kolkata.

Acknowledgments

Calcutta Book fair is very attractive to me and I go there every year to purchase some books from an enormous display of books for sale. This book fair created an urge in my mind to write books that may be displayed in the fair. I could not imagine my dream would become a reality. My first book titled "Grahantore Manush", roughly translated as "People on a distant Planet", dealing with Science and Yoga at the fictional plane, has been appreciated by children beyond my expectation. I sent a copy of this book and its translation to Professor Stephen Hawking for his opinion and waited for a few months with a hope of his reply. At last one day I received the most coveted reply dated 23 March 2006. He remarked that 'Thank you for your translation of the book and your imaginative story for Children'.

Professor Hawking is an inspiration for all the people of our planet and I always get inspiration from his books, life and the letter he sent me.

He is no more, we lost him on 14th March, 2018 but he will remain in his books and in the bottom of our hearts forever.

I am fortunate enough to have renowned persons as my well-wishers. Dr. Debiprosad Duari, Director, M.P. Birla Institute of Fundamental Research, popularly known as M.P. Birla Planetarium, advised me to write my next book in English. My hesitation in writing in English disappeared by words of encouragement bestowed upon me by my well-wishers. Dr. Pradip Kumar Majumdar, Professor and the author of many mathematics books, has guided me in writing this book.

I have had a lot of help with this book from Mrs. Shyamala Chatterjee, my son's English teacher. She has guided me in writing correct English.

I have listed the books in the references that have been indispensable to me in the writing of this book. I have the greatest respect for the authors.

The help and support I have received from my wife, Mrs. Bani Banerjee, my son Soumyadip Banerjee and his computer teacher Rajesh Mondal, my niece Paramita Chatterjee and her husband Anish Sarkar, have made it possible for me to make my manuscript suitable for printing.

My office colleagues Shyamaprasad Mukherjee, Senani Das and Asit Baran Bhowmik have encouraged me in my writing.

I have had suggestions of how to improve the book from my uncle Shakti Mukherjee and friends Dr.Goutam Pathak, Netai Das, Bikash Hore, Swapan Biswas, Ardhendu Karmakar, Tapan Aru and my cousins Debdas Banerjee, Sanjoy Mukherjee, Somnath Mukherjee.

I am grateful to Professor Tapan Kumar Banerjee, my former Professor of Mathematics at Serampore College, for writing the preface and editing this book.

I am very grateful to the publisher for publishing Ramanujan & Ancient Indian Mathematicians in English.

Swapan Banerjee

PREFACE

Srinivasa Ramanujan Iyengar, one of the world's greatest mathematical geniuses, was born in India. Nowadays he is respected all over the world for his talent. He made substantial contributions to the analytical theory of numbers and worked on elliptic functions, continued fractions and infinite series.

Ramunujan came of a very poor family and was not in a condition to bloom with health and happiness. Poverty could not prevent him from progressing. Ramanujan's patience, perseverance and passion for mathematics reached him to the Cambridge University.

Professor Godfrey Harold Hardy first identified the talent of Ramanujan from a letter sent to him by the genius. G.H. Hardy endeavored to make him a triumphant mathematician. Professor Hardy, while rating geniuses on a scale of 100, put most of them in the range of around 30, giving a rating of 60 to the rare exception. However, for Ramanujan, he suggested, only the value of 100 would fit.

In ancient time people of Rome, Greece, Babylon, China, Egypt, India and many other countries had their own numerals to write numbers. Ancient Indian mathematicians evolved a numeral system which we call – The Decimal System. It was a creative endeavor on the basis with only ten symbols (1, 2, 3, 4, 5, 6, 7, 8, 9 and 0) – they should be able to represent any number as well as having their individual value.

Our ancient mathematicians and astronomers namely Baudhayana, Aryabhata I, Aryabhata II, Bhaskara I, Bhaskara II, Varahamihira, Brahmagupta, Sridharacharya, Madhava, Nilakantha

Somayaji and many others moulded modern mathematics which is indispensable even today.

The work of ancient mathematicians of India provides the context of Albert Einstein's remark that 'We owe a lot to the Indians who taught us how to count, without which no worthwhile scientific discovery could have been made.'

This book is a lucid and elegant expression of Ramanujan's biography and ancient Indian mathematicians. Writer Swapan Banerjee deserves our sincere thanks for his endeavor.

Tapan Kumar Banerjee
Former Professor of Mathematics,
Serampore College, Hooghly, West Bengal

Life
Of
Ramanujan

1. Life Of Ramanujan

I used to come home on Saturdays and set out for my working place every Monday morning. That Monday morning was severely cold due to inclement weather and the wind was biting. My elusive memory cannot remember the date but the month was December of 1987.

I have a habit of reading The Statesman and Science Reporter in the train. Newspaper vendors on the platform supplied me with the newspaper and Science Reporter of December issue. I had never seen the picture of the person printed on the cover page of Science Reporter, the name printed below the picture was also unknown to me.

The train was approaching the platform, I was ready to board the train. Luckily I got a seat by the side of a window and began to read the Science Reporter.

The great genius mathematician was born on December 22, 1887 in India, but it is a pity that I had not heard his name before. December 22, 1987 was his birth centenary. So Science Reporter printed his picture on the cover page and some articles about him in the magazine itself. I read the articles attentively in the train; the articles made me proud that he was an Indian, it also plunged me in deep sorrow when I came to know that we lost a rare genius like him when he was only thirty two years and four months old.

A person who could achieve a position among the great mathematicians of the world at only thirty two years of age could have become the greatest mathematician in the world, had he lived for another thirty years. Geniuses like Rabindranath Tagore, C.V.Raman and many others were born in India but mathematical genius Srinivasa Ramanujan is beyond comparison or praise.

We know that diamond cuts diamond. The great mathematician G.H.Hardy could recognize the genius in Ramanujan. G.H.Hardy once said, "Ramanujan was my discovery. I did not invent him-like other great men, he invented himself-but I was the first really competent person who had the chance to see some of his work, and I can still remember with satisfaction that I could recognize at once what a treasure I had found." Ramanujan wrote several letters to famous mathematicians of England but only G.H.Hardy recognized the real diamond in Ramanujan.

G.H.Hardy

I could have written a few lines to give an idea about the talent of Ramanujan, but it is not possible for a petty writer like me to do the same. It would be better if I insert a quotation from the book 'Ignited Minds' written by our former president and renowned scientist Dr. A.P.J. Abdul Kalam to express an idea about the talent of Ramanujan.

"Then comes to my mind the greatest of all geniuses ever known and acknowledged, and who lived within our present memory- Srinivasa Ramanujan. He lived only thirty three years (1987-1920) and had no practical formal education or means of living .Yet, his inexhaustible spirit and love for his subject enabled him to make a vast contribution to mathematical research and some of his contributions are still under serious study, engaging the efforts of mathematicians to establish formal proofs. Ramanujan was a unique Indian genius who could melt the heart of as rigorous a mathematician as Prof. G.H.Hardy of Trinity College, Cambridge. In fact, it is not an exaggeration to say that it was Hardy who discovered Ramanujan for the world. Why do not our reputed scientists locate another Ramanujan in our schools? Oh my friends why don't you in every field integrate and grow instead of differentiating!

'Every integer is a personal friend of Ramanujan', one of the tributes to Ramanujan said and it was no exaggeration. Prof. Hardy, while rating geniuses on a scale of 100, put most of them in the range of around 30, giving a rating of 60 to the rare exception. However, for Ramanujan, he suggested, only the value of 100 would fit. There can be no better tribute to either Ramanujan or to the Indian heritage. Ramanujan's work covers vast areas including prime numbers, hyper geometric series, modular functions, elliptic functions, mock theta functions, even magic squares, apart from some serious work on the geometry of ellipses, squaring the circle and so on."

1.1 Childhood of Ramanujan

Komalatammal

The ancestral home of Ramanujan is at Kumbakonam. This town is situated about 160 miles south of Chennai (previous name Madras) and on the bank of the Cauvery, the Ganges of the South.

Komalatammal, mother of Ramanujan went to Erode, her parental home, two months before her first child was to be born. Erode is also situated on the bank of the Cauvery, at a distance of 250 miles South West of Madras.

Just after sunset was born a bright and genius son whose fame spread like the rays of the sun in later life. That auspicious day was Thursday, December 22, 1887.

On the eleventh day the child was christened Srinivasa Ramanujan Iyengar. His father was Srinivasa Iyengar. According to south Indian tradition the father's name is bestowed before the son's name. Iyengar is a caste name for a branch of South Indian Brahmins. The particular name of the child became Ramanujan.

A Vaishnavite saint named Ramanuja lived in India around 11th Century A.D. He was a great spiritual reformer and he rejuvenated Hinduism which was in low esteem. He was also born on Thursday, and had some astrological identity with Ramanujan, thus Ramanujan got his name from this great saint.

India has two great epics : the Ramayana and the Mahabharata. The Mahabharata is the longest epic in the world. In the Ramayana lord Ram is the ideal of Indian manhood and Lakshman is his ideal younger brother. Here anuja means younger brother and Ramanuja means younger brother of Ram. Just after one year of birth Ramanujan came to his ancestral home at Kumbakonam with his mother.

The father of Ramanujan was a clerk in a cloth shop; he was good at appraising fabrics. His salary of Rs. 20 per month was not enough to maintain the family properly. Mother Komalatammal used to sing bhajans or devotional songs at a nearby temple in a group. Half of the income from these performances went to the temple, the other half was distributed among the singers. In this way she earned 5 to 10 rupees per month which was enough for a poverty stricken family. She never absented herself from the devotional songs.

Ramanujan's house on Sarangapani Sannidhi Street

Ramanujan's mother was not attending the devotional songs program for four or five days in December 1889. One day the head of the singing group visited Komalatammal's house to enquire about her absence. Entering the room, she saw the baby Ramanujan lying on a bed made of margosa (neem) leaves. Komalatammal was rubbing gently margosa leaves mixed with turmeric (haldi) while chanting devotional songs.

Baby Ramanujan had smallpox and his affectionate mother was trying to give some relief from itching and fever with margosa leaves and turmeric. Ramanujan bore the scars of smallpox all his life. But we are lucky enough that though four thousand people died from this disease at that time Ramanujan came out of danger.

At that time, arrangements for treatment of diseases in India was insignificant. Moreover investigation into the causes of disease and discovery of medicines in those days i.e.125 years ago, were unlike now a days. Ramanujan lost one sister and two brothers at a tender age, as child mortality rate was very high then.

A brother of Ramanujan was born in 1898, he was named Lakshmi Narasinham. When Ramanujan was seventeen, another brother Tirunarayanan was born. Both of them remained alive and lived long.

He was so stubborn in his childhood that he would wallow in mud if he was not provided with his favorite food. He remained speechless up to the age of three and neighbors thought that the child would be dumb. His mother visited her parental home frequently. His grandfather endeavored to make him talk and taught him Tamil alphabets and he was successful.

On October 1, 1892, he was enrolled to a local *pial* school. In South India a *pial* school was usually conducted on some veranda of a house. A teacher with six or seven pupils was a common picture in those schools. Their method of teaching and teachers were both unattractive to him. He was so adamant at that tender age that none could persuade him to go to school most of the days. He was enrolled to various schools in his native town and maternal uncle's town Kanchipuram; he was even admitted to a school in Madras where his grandfather was working. But all efforts were in vain. The child Ramanujan found no teacher or his teaching attractive in any school. At last the parents of Ramanujan took the help of a local constable to scare him into going to school.

The Child Ramanujan had no interest in sports and games, unlike other children. Lack of exercise made Ramanajun obese. Other children of his age were afraid of quarreling or fighting with him, for if he fell on any one of them, the child would be crushed.

Upanayanam or sacred thread ceremony was performed when he was only five years old. After this ceremony he became a confirmed Brahmin and acquired the rights to perform rites of Hinduism.

Ramanujan's thatched house was situated beside Sarangapani Sannidhi Street. The street was thirty feet wide and was ideal for the playing of local children in the afternoon. Gradually when the child Ramanujan became attentive to his studies, he used to engage himself in mathematics practice on a slate in their veranda while other children were playing on the road in front of him. From that tender age, he became so devoted to mathematics that when children of his age were playing in front of him, he was indifferent to that kind of enjoyment.

At the age of ten in the month of November of 1897, he stood first in his district in the final examination of primary school. He got very good marks in all subjects. His success brought glory to his school, Kangayan Primary School. He got admission to Town High School in January 1898.

At that time Kumbakonam town was famous for fine silk sarees. There were about two thousand weavers' looms in the town, it was also famous for fine metalwork. About six hundred craftsmen lived there and their handiwork was executed in copper, silver and brass. There was great demand for the silk and metal products in aristocratic families of Europe.

As Ramanujan's father used to be engaged in his job from dawn to dusk, Ramanujan was brought up mainly by the guidance of his mother.

Town High School was a five - minute walk from his house. Brilliant students got a chance to study in that school and they prospered in their later life. While Ramanujan was a student of that school S.Krishnaswami Iyer was the headmaster of that school, he was incumbent in that position for twenty two years. The gravity of his demeanor in the school was awful to the students. He used to stroll among classes with a walking stick in hand; sometimes he would enter a class and begin to teach with much vigor, keeping the teacher aside, but his teaching was lively.

Town High School

S.Krishnaswami Iyer, the headmaster first noticed the mathematical genius in Ramanujan. The classmates of Ramanujan sought his help to solve mathematical problems. Even students from higher classes took his help to solve difficult sums.

The poverty-stricken family of Ramanujan arranged for two Brahmin students to stay in their house as paying guests. Thus Ramanujan got two elder brothers for helping him in his studies. The excessive zeal of Ramanujan compelled college students to bestow their whole knowledge of mathematics to Ramanujan. After acquiring knowledge from them, he began to pester them for bringing books of mathematics from the college library. They brought many books one after another and S.L.Loney's Trigonometry was one of them. That book was beyond comprehension for a boy of thirteen years, but Ramanujan comprehended the book and solved all the problems in the book at that age.

Students senior to him by two or three years, used to come to him to get difficult problems solved and Ramanujan worked out the sums within a few minutes. Even his mathematics teacher, Ganapathi Subbier, the senior mathematics teacher of Town High School used to sought his beloved student's help to solve difficult problems. Sometimes his questions in mathematics classes were embarrassing to teachers, as they could hardly answer them.

One day, one of his mathematics teachers taught the students, "If a number is divided by the same number the result is one. If three fruits are divided among three persons everyone gets one fruit. If one thousand fruits are divided among one thousand people every one gets one fruit, the result is the same that is one". Ramanujan did not agree with his teacher, he stood up and said, "Is zero divided by zero also one? If no fruits are divided among no one, will each still get one?" He became a minor celebrity among the students of his school.

His performance brought a brilliant result in his school final examination in 1904. He received K. Ranganatha Rao award for excellence in Mathematics. In the prize distribution ceremony, S. Krishnaswami Iyer, the headmaster said "Hundred marks out of hundred is not the proper evaluation for Ramanujan".

1.2 College Student Ramanujan

As Ramanujan was the best student in school he got admission in Kumbakonam Government College in 1904 with a scholarship. In college, arts subjects became obstacles to his progress. In school he got full marks in mathematics and passed other subjects with good marks with little effort. But in college, arts subjects required more attention as they were more difficult than in the school curriculum. Due to his tremendous attraction towards his favorite subject mathematics, he gradually became weak in other subjects.

Perhaps through college students who were living with his family or by other means Ramanujan got a book named 'A Synopsis of Elementary Results in Pure and Applied Mathematics' in 1903. The book and its writer were not worthy of mention, but the book and its writer George Shoobridge Carr became famous as they got a permanent place in Ramanujan's biography. Carr was a mathematician who for many years tutored privately the Tripos examinees in London. Ranking in this examination was just like crossing the threshold to a career. The book was a collection of his coaching notes and was popular among Tripos examinees. Ramanujan got the first volume of that book and it was full of theorems, formulas, geometric diagrams, including subjects like algebra, trigonometry, calculus, analytic geometry, differential equations etc.

Ramanujan's fun and delight were to work out difficult problems. He had no money to purchase paper but that was not a problem for him. He had his big slate; he solved problems and wiped the writing out with his elbow. He was not conscious about time, when studying, midnight had passed and dawn would peep through the window.

He rediscovered many formulas invented by European Mathematicians a hundred years ago as he was not acquainted with modern mathematics. He remained indifferent to all his classes but his alertness could be measured in mathematics class by his shining bright eyes. Though he scored brilliant marks in mathematics, he failed in English composition in F. A. examination. This failure became an obstacle to his career, as the scholarship was stopped. It became very difficult for poor Ramanujan to continue his study. His mother with an anguished heart, met the principal. "How could you refuse my son a scholarship? He excels in mathematics." The principal behaved politely with his mother but he was firm about rules. Ramanujan's mother could not budge him.

The meager income of his father was inadequate for continuing Ramanujan's study after meeting the expenses of a family of seven persons. Ramanujan vowed to study all subjects attentively, allotting less time to his favorite subject. Ramanujan was intelligent and he knew very well that they had been living precariously on the income of his father's meager salary. So he started to study all subjects heart and soul, allotting less time to mathematics. He vowed to pass next year as well as to revive his coveted scholarship.

It was a pity for him but boon to the mathematical world that he could not apply his mind to arts subjects. Ramanujan was a mathematical genius; his instinct was to carry on research in mathematics. His passion for mathematics always drove his mind towards his favorite subject. Next time in 1905 he sat for the F.A (First Arts) examination .The result was the same .He got full marks in mathematics, but poor marks in other subjects. Ramanujan could not bear this agony, he ran away from home in August 1905, with an unsound mind. He boarded a train.

The train was introduced in India by the British in the mid-nineteenth century. The inaugural train in India ran from Bombay (Mumbai) to Thane on 16 April, 1853.In undivided Bengal the first train ran from Howrah to Hooghly on 15 August, 1854. Railway brought remote places of India to proximity. It also banished untouchability in a train compartment where all castes of Hindu and people of other religions were compelled to sit, touching each other in a crowded coach, though they resumed their superstition outside the station.

Ramanujan reached Vizagapatnam (Vizag). Perhaps he went there to earn money which was a crying need for him as well as for his family. Being unsuccessful in finding a livelihood there, he returned from Vizagapatnam. One day he fell fast asleep in railway waiting room of Egmore station. A noble person woke him up and took him to his house. He treated Ramanujan cordially at home and sent him to Pachaiyappa's College of Madras.

In this new college he started with renewed vigor and enthusiasm. Ramanujan showed his notebooks to his new math teacher. He was so impressed that he introduced Ramanujan to the principal. The principal realized that Ramanujan was not an ordinary student but a prodigy. He awarded him a partial scholarship to lessen his pecuniary difficulties.

He started to live in his grandmother's house which was situated in a small lane off the fruit market. Light and air were afraid to enter into this house which was in a dark and filthy area. He was optimistic in the first year though he lost three months due to a bad bout of dysentery. The disease compelled him to go back home in Kumbakonam.

Ramanujan was in the good books of his math teacher N.Ramanujachariar. After working out a problem on the blackboard he would ask, "What's your opinion Ramanujan?" Ramanujan did not prefer writing down several steps to solve a problem. He used to solve a problem on the blackboard in three or four steps sometimes with a different method, leaving his classmates confused and his professor proud.

P. Singaravelu Mudaliar, senior math professor of his college became his intimate friend. They used to solve problems appearing in mathematical journals. A problem unsolved by Ramanujan would be given to the professor to work out at home. Invariably the professor also could not solve it.

In Pachaiyappa's College he had to read a new text book, Physiology for Beginners. It was boring and intricate to him. Ramanujan appeared for the F.A. examination in December 1906 and failed. He got poor marks in all subjects except mathematics. He took only thirty minutes from the three-hour math examination, solving many problems with various methods. He scored less than ten percent marks in his new boring subject Physiology. Next year he sat for the same examination and the result was the same. He became unsuccessful four times in F.A. examination in a row; twice from Government College, Kumbakonam in 1904 and 1905 and twice from Pachaiyappa's College in 1906 and 1907. All doors to higher education for Ramanujan were shut, though he was a marvel in mathematics .That was the education system in India which would not budge for a genius.

1.3 Unemployed Ramanujan

Ramanujan discovered many formulas even when he was a school student. His discoveries were noted down by him in his note book. The number of notebooks increased day by day, and became a priceless or invaluable treasure of three to four thousand theorems. Ramanujan had even no money to purchase paper, so he had to depend on his big slate for his mathematical research. Sometimes they had to starve for lack of money. Illiterate aunts of his locality had no knowledge of mathematics but they could understand that Ramanujan had been engaging in an austere endeavor to achieve something eternal. Yes, Ramanujan got absorbed in an ascetic striving to express thoughts of god through mathematics. Once he told a friend "An equation for me has no meaning unless it expresses a thought of god." His aunts sent food to his home or escort Ramanujan to their houses to feed him. There was no aunt who would deprive Ramanujan of her affection.

He could not finish college education, so it was difficult to get a job. Though he was a genius in mathematics he remained unsuccessful in finding tuition in his favorite subject. He could not confine himself to the course material of his students. He used to cross the boundary of his students' syllabus, which was beyond the comprehension of his students and unnecessary for passing examinations.

Viswanatha Sastri was the son of a professor of philosophy of a Government college. Viswanatha's father engaged Ramanujan to coach his son. Early in the morning Ramanujan had to walk a long distance to teach his student algebra, geometry, trigonometry for seven rupees a month. Viswanatha found in Ramanujan an inspiring teacher but of a higher stratum.

N.Govindaraja Iyengar, was a classmate of Ramanujan who asked him to help with differential calculus for his B.A. examination. This arrangement lasted only for two weeks. In later life Govindaraja became chairman of India's public service commission. I quote from Govindaraja. "He would talk only of infinity. I felt that his tuition would not help me in the examination, so I discontinued it".

Ramanujan had no money and found no source to earn money. At home he remained absorbed in mathematics or in deep thought. Neighbors of Komalatammal advised her to marry her son to a suitable bride. After marriage he would be inclined to lead a domestic life and consequently, try to get a job. Village Rajendram is situated about sixty miles west of Kumbakonam. One day Komalatammal visited her friends in that village in search of a bride. She chose a bright eyed daughter of a distant relative, named Janaki. Only choosing a bride would not suffice for a marriage at least one hundred years ago. She drew her son's horoscope and compared it to that of Janaki. Ramanujan's mother concluded a good match according to horoscopes.

Nothing was smooth sailing throughout Ramanujan's life, even at the time of marriage. He set off with his mother for Rajendram by a train .The train was abnormally late. The groom was not coming and auspicious time for marriage was passing out, so Janaki's father became very anxious. He began to contemplate seeking another groom to marry her daughter that night .At that time there was no phone or mobile. Telegraph was introduced in India in 1839 and its use was very limited. This old and very essential system of communication has been discontinued in India from 15.07.2013. At last he came before dawn by a bullock cart after getting down from the train. Every one present there became happy as the groom came at last. Rituals of marriage were completed, that day, on 14 July 1909. Then Janaki was nine years old.

Responsibility increased, money was needed for maintenance of both, Ramanujan and Janaki. Any type of job anywhere in India was badly required, but who would provide him a job? The person, who should be deeply absorbed in mathematical research, was frantically searching for a job. Simple food was needed for mathematical research in a cool brain. He had no B.A degree so he used to carry his note books while searching for a job. He showed his notebooks to employers but they could not understand his work. He had been suffering from hydrocele but had no money for operation; even friends and relatives did not come forward to help. At last, Dr. Kappuswami operated without taking any fees in January, 1910. He continued his efforts in search of a job. One day a famous professor of mathematics said "Though you are good in mathematics, you have no B.A. degree and so you are worthless." In those days a person who could not complete his school education had not to face much difficulty to get a job. Ramanujan completed school education with a brilliant result; nevertheless he was unemployed for several years.

Ramanujan came to Tirukoilur town in the second half of 1910. V.Ramaswami Iyer was then deputy collector of that area as well as a renowned mathematician. Ramanujan came here to meet him as he had recently founded the Indian Mathematical Society. He met him with a hope in mind that he would provide him with employment. Ramanujan showed him his notebooks and the unfamiliar mathematics astonished Mr. Iyer. He had no intention of suppressing the genius by engaging him in a clerical job in the revenue department. So he wrote a letter of introduction and sent

Ramanujan to P.V. Seshu Iyer his mathematician friend in Madras.

P.V. Seshu Iyer was a professor of Government College in Kumbakonam while Ramanujan was a student in that college. Now, after four years, P.V.Seshu Iyer was in Presidency College in Madras. The professor and his student met after four years. He was acquainted with his student's talent. He also sent Ramanujan to S. Balakrishna Iyer a mathematics lecturer at Teachers' College in the Madras suburb, with a letter of introduction. Mr. Balakrishna was a gentle person. He offered him coffee and looked through his notebooks, which were beyond his comprehension. He urged his English boss, Mr. Dodwell, four times on behalf of Ramanujan for a clerical job at any poor salary in his college. But Ramanujan was so unlucky that he got no job, however poorly it might be paid.

P.V. Seshu Iyer

In December Ramanujan came again to Madras in search of a job. He needed an employment to survive; the amount of salary was not a factor. All of a sudden, he met C.V.Rajagopalachari, a renowned lawyer, who was few months older than him and they came from the same town, the same school and used to visit the same temple for religious functions. Now he was a lawyer and Ramanujan was an unfortunate unemployed vagabond. Rajagopalachari proved himself really; a friend in need and a friend in deed. He cordially received him and arranged for his food and shelter for a few days. He also assured him of a return ticket to Kumbakonam as his friend had no money.

One day during recess, a math duel took place between two students. A clever proud student of a higher class handed

Ramanujan a math problem, which is mentioned below:-

$$\sqrt{x} + y = 7$$
$$\sqrt{y} + x = 11$$

The student thought that Ramanujan would not be able to solve it and he would not remain clever in the school. Seeing the problem all students present there got upset. Ramanujan took thirty seconds only to solve it mentally; the solution is x = 9 and y = 4. This incident of 1902 influenced Rajagopalachari and made their friendship very strong.

Depressed Ramanujan expressed his plight to his friend. He was unsuccessful in colleges, none appreciated his work and he had sent samples of his work to the Indian Mathematical Society and to famous professor Saldhana of Bombay. He received no response from them till that day. His future was bleak.

One day Ramanujan accompanied Rajagopalachari to meet R.Ramachandra Rao. He was district collector of Nellore as well as a mathematician. He was also holding the post of secretary of the Indian Mathematical Society founded by V.Ramaswami Iyer. He was wealthy and influential. He had been named Dewan Bahadur. Ramanujan anticipated a lot from him. The nephew of Dewan Bahadur R.Krishna Rao introduced Ramanujan with him. At first sight R.Ramachandra Rao saw a person before him, short in figure, unshaved, untidy indicating indigence. But he was stout with conspicuous or distinguished bright shining eyes and a note book under his arm.

R.Ramachandra Rao

On that day R.Ramachandra Rao asked Ramanujan to keep his note book for a few days. When Ramanujan met him the second time, he returned the note book to Ramanujan as he had not seen that type of theorem before and it was not convincing or understandable to him. Third time when Ramanujan and his friend met him he refused Ramanujan as he had doubts about the credibility of Ramanujan. While they were returning Ramanujan told C.V. Rajagopalachari, his friend, about professor Saldhana's letter which he had received recently.

Professor Saldhana mentioned the subtlety of Ramanujan's theorems in his reply letter .Then and there they returned to meet R.Ramachandra Rao, the fourth time, as C.V.Rajagopalachari, his friend was determined that Ramanujan should get a job. He became angry "Here again just a few minutes later?" was his expression. He saw Professor Saldhana's reply, and he was shown some of Ramanujan's easier theorems. Doubts disappeared from his mind. R.Ramachandra Rao could clearly perceive that Ramanujan was a man of rare talent. In his words "Then step by step he led me to elliptic integrals, and hyper geometric series. At last, his theory of divergent series, not yet announced to the world, converted me. I asked him what he wanted." He wanted to earn a small amount of money or scholarship to survive for research work in mathematics. R.Ramachandra Rao narrated his desire later. "He wanted leisure, in other words, simple food to be provided to him without exertion on his part, and that he should be allowed to dream on."

R.Ramachandra Rao was a judicious person. He could provide him with a job in his office in Nellore but he did not want to suppress Ramanujan's talent with the burden of a petty clerical job. He tried his level best for a scholarship for Ramanujan but became unsuccessful. Now, he started to send him a monthly donation of rupees twenty five only, from his earnings by money order.

Now, Ramanujan began to stay in Madras from early 1911,

for the next three years. In May 1911, he shifted to a boarding house named Summer House on Swami Pillai Street in Madras. The boarding house was not up to the standard of its gorgeous name, but suitable for Ramanujan to continue his mathematics with other students.

Though he continued his mathematics without any anxiety for money, he remained in constant mental agony as he was taking financial help from a person without doing any work for him. The money he was receiving was adequate for his food and lodging in the boarding house but he could not afford paper for working his mathematics. He required four reams of paper every month; who would provide that?

Ramanujan's Slate

His large slate was his companion in his poverty. He remained thoughtful with mathematics all the time. It was a waste of time for him to pick a rag, so he used to write and erase by elbow to make the process quick. His elbow became rough, dirty and black.1911 was a remarkable year in the history of India. The capital of India was shifted from Calcutta (now Kolkata) to Delhi with great fervor. And that was the year in which Ramanujan's paper first appeared in the Journal of the Indian Mathematical Society. Thus he got the opportunity to be acquainted with Mathematicians of India and abroad. The few papers that appeared in that journal in that year were appreciated by mathematicians. The publication of these papers gave delight to him but that could not wipe out his mental agony as he was dependent on dole.

1.4 Self-Reliant Ramanujan

1912 became an auspicious year for Ramanujan. He got a temporary job in the Madras Accountant General's office. The salary was a mere twenty rupees per month but was enough to restore confidence in Ramanujan as he would not have to depend on R. Ramachandra Rao's generosity which had continued for one year. He served there from 12 January to 21 February.

According to the advice of his well-wishers, Ramanujan applied for a permanent job to the Chief Accountant of Madras Port Trust on 9th February. The application is appended below.

From   9th February 1912
S. Ramanujan   Triplicane
7, Summer House
Triplicane

To
The Chief Accountant
Port Trust
Madras

Sir,

I understand there is a clerkship vacant in your office and I beg to apply for the same. I have passed the matriculation examination and studied up to the F.A. but was prevented from pursuing my studies further owing to several untoward circumstances. I have, however, been devoting all my time to mathematics and developing the subject. I can say I am quite confident I can do justice to my work if I am appointed to the post. I therefore beg to request that you will be good enough to confer the appointment on me.

I beg to remain, Sir,
Your most obedient servant,
S.Ramanujan

(Ramanujan was residing at summer house in Triplicane area of Madras when he applied).

He was selected on 25th February and joined the service on 1st March as a class III, Grade IV Clerk. He had two superiors; the first person was Sir Francis Spring, the chairman of Madras Port Trust and the second person was Narayana Iyer, the office manager, both were noble mathematicians. Both his superiors played an important role in blooming his talent fully.

His earnings started at thirty rupees per month. His joy knew no bounds as he would not have to worry for food and would be able to devote himself to mathematics. Except during office hours, he used to keep himself engaged in mathematics. The word 'waste' was not written in his dictionary. In the morning he kept himself engaged in mathematics up to the moment of departure to office .In the evening after returning from office he absorbed himself in his favorite subject.

Until now his wife Janaki used to frequent her father's house and her father- in-law's house. Now she was brought to Madras to live permanently with Ramanujan.

In the morning it was an usual sight for his friends and

others that he was running along the Beach Road for going to office with his coat, kutumi (tuft of hair) unmanageable in the breeze. Service in Port Trust brought emancipation of Ramanujan; it also minimized dearth of paper. At tiffin break he used to loiter around the dock for collecting packing paper. He required four reams of paper per month and it was not possible for him to purchase that amount of paper from his salary. So packing paper was a boon to him.

His wife Janaki Devi later recalled that her husband used to start his mathematical research just after returning from office. His perseverance seemed to be endless. Most of the days he continued his work till mid-night. On some days, the rising sun peeped through the window, after which he slept for two or three hours, ate some food and ran to his office.

Narayana Iyer was known to him much before joining Port Trust Office. They remained engaged in their mathematics culture at Narayana's house off and on; that also continued up to not less than mid-night.

Narayana Iyer

Narayana Iyer urged Ramanujan several times to write down more steps while solving problems. If a problem requires at least fifteen steps to solve, Ramanujan would solve that within three or four steps. Narayana Iyer told Ramanujan that what was not convincing to him, his close associate, how could he convince others? Ramanujan's brain was more sophisticated than a modern computer, so he did not want to waste time by writing more steps.

Perhaps Ramanujan's brain remained active during his sleep. He would wake up anytime during sleep and record a solution that had come to his mind, in dream, according to him. The hurricane lamp was kept dim for his sudden activity. This information was obtained from Narayana Iyer's son N. Subbanarayanan.

Narayana Iyer was not only a boss but also a mentor and friend of Ramanujan. He convinced Sir Francis Spring about the talent of Ramanujan. Sir Francis also became a well-wisher of Ramanujan.

One day Sir Francis called Narayana Iyer into his office chamber. He asked him grimly "Why have you kept these mathematical papers in this important file?"

Narayana Iyer replied, "I am innocent, it is not my handwriting. It is Ramanujan's work". Sir Francis laughed and did not say anything about it to Ramanujan.

Narayana Iyer used to say that everyone thought Ramanujan was ordinary glass but one day everyone would come to know that he was a diamond.

The Journal of Indian Mathematical Society had been publishing his papers but the response was not up to the mark. Then perhaps in India there was no such mathematician who could realize the importance of Ramanujan's inventions.

So with the help of Sir Francis Spring and Narayana Iyer he drafted a letter including samples of his work. He sent the letter to H.F. Baker, the renowned mathematician who was an FRS and the president of London Mathematical Society. There was no response. Ramanujan wrote another letter to E.W. Hobson an FRS and an eminent mathematician and an incumbent of Sadlerian Chair in pure mathematics of Cambridge University. There was no reply.

How could a clerk, who could not pass college examination several times, invent any mathematical formula or theorem ? So the two celebrities of mathematics did not feel the urge to respond. The amount of publicity that they got for not responding to Ramanujan was more than their profession as a mathematician, for they got a permanent place in Ramanujan's biography.

He was disappointed but not depressed. He wrote another letter with much vigor on January 16, 1913 and sent it to G.H.Hardy, a young mathematician of Cambridge University. Yes, this time he had written to the right person. G.H.Hardy brought a miracle of transformation in Ramanujan's life. The incidents that occurred onwards were more thrilling than a thriller film. The letter ushered in a new era in the history of mathematics. The letter is appended below.

16th January
Madras

Dear Sir,

I beg to introduce myself to you as a clerk in the Accounts Department of the Port Trust Office at Madras on a salary of only £20 per annum. I am now about 23 years of age. I have had no University education but I have undergone the ordinary school course. After leaving school I have been employing the spare time at my disposal to work at mathematics. I have not trodden through the conventional regular course which is followed in a University course, but I am striking out a new path for myself. I have made a special investigation of divergent series in general and the results I get are termed by the local mathematicians as "startling".

Just as in elementary mathematics you give a meaning to a^n when n is negative and fractional to conform to the law which holds when n is a positive integer, similarly the whole of my investigations proceed on giving a meaning to Eulerian Second Integral for all values of n. My friends who have gone through the regular course of University education tell me that $\int_0^\infty x^{n-1} e^{-x} dx = \Gamma(n)$

is true only when n is positive. They say that this integral relation is not true when n is negative. Supposing this is true only for positive values of n and also supposing the definition $n\Gamma(n) = \Gamma(n+1)$ to be universally true, I have given meanings to these integrals and under the conditions I state the integral is true for all values of n negative and fractional. My whole investigations are based upon this and I have been developing this to a remarkable extent so much so that the local mathematicians are not able to understand me in my higher flights.

Very recently I came across a tract published by you styled Orders of Infinity in page 36 of which I find a statement that no definite expression has been as yet found for the number of prime numbers less than any given number. I have found an expression which very nearly approximates to the real result, the error being negligible. I would request you to go through the enclosed papers. Being poor, if you are convinced that there is anything of value I would like to have my theorems published. I have not given the actual investigations nor the expressions that I get but I have indicated the lines on which I proceed. Being inexperienced I would very highly value any advice you give me. Requesting to be excused for the trouble I give you.

I remain,
Dear Sir,
Yours truly,
S.Ramanujan.

In those days letters were not dispatched by aeroplanes, ships were used for that purpose. Aeroplane was invented by Orville Wright and Wilbur Wright on 17 December 1903. Flights between countries began much later. So the letter reached Cambridge at the end of January.

G.H. Hardy had a habit of reading the newspaper London Times during breakfast. On one fine morning, during breakfast G.H. Hardy received Ramanujan's thick letter of ten pages. He read the letter and had a superficial look on the samples of mathematical results. He could not comprehend Ramanujan's mathematics as they were not proved following a process much in vogue. He could not decide within that short time whether that was a letter from an eccentric person or a genius from India .He kept aside the letter and engaged himself in his daily routine.

G.H. Hardy was then thirty five years of age, quite thin and looked younger than his age. He could easily mingle with his students. He remained a bachelor all his life. Though he was an English man he never drank alcohol and was free from inglorious habits. He was free from the slightest hesitation in accepting a person of dark skin if he was a man of qualities .He was a magnanimous person and free from all prejudices. In those days Britain was not liberal regarding woman's education. Women were not awarded degrees from any university, though they deserved them according to their merit. Many women got university degrees due to the endeavors of G.H.Hardy at that time.

In this context I may add a few lines regarding women's education in India. Raja Ram Mohan Ray tried his best to abolish the practice of widows burning themselves on their husbands' pyres. He succeeded and Lord William Bentinck prohibited 'sati' in 1829. Another pioneer from Bengal Iswar Chandra Vidyasagar who established many girls' schools in undivided Bengal. Educationist

Bethune established girls' schools in Calcutta (now Kolkata). After coming to India Sister Nivedita took a pioneering role in Women's education at that time. I feel proud to write that our Bengali women Kadambini Bose and Chandramukhi Basu got degrees in 1883. They were the first women in the entire British Empire to get degrees from a university i.e. no British woman got degrees before them.

Though G. H. Hardy kept aside the letter in the morning, it haunted him throughout the day during his work. He sent a small letter to his friend and colleague John Edensor Littlewood in which he wished to meet him after dinner. They were bosom friends and their papers were published jointly in journals.

Littlewood was senior to Ramanujan only by two years. Littlewood was an established mathematician but so long Ramanujan got no recognition though he had merit and capability.

G.H. Hardy spread out Ramanujan's letter, formulas etc. before Littlewood. They started from 9P.M in the severe winter of January. They pointed out that some formulas were rediscovered by Ramanujan, but he had no idea about the discovery of Laplace and Jacobi a hundred or one hundred and fifty years ago. They could prove some of the theorems and some were unfamiliar to them. After three hours of endeavor they came to the

John Edensor Littlewood

conclusion that the letter and the samples of theorems were sent by a mathematical genius.

Hardy and Littlewood smiled at each other in joy. The joy was not in inventing a formula in mathematics but discovering a mathematical genius like Newton.

The incidents that took place onwards were just like an exciting cinema. Before sending his reply, G.H. Hardy informed about Ramanujan's talent to several mathematicians of England. He also requested the mathematicians in India, who were acquainted with him and holding higher posts, that Ramanujan was a mathematical genius and all arrangements should be made for his research.

In reply to Ramanujan's letter G.H.Hardy wrote that he had been interested by his theorems but he had not sent convincing proofs. So it had become difficult for him to access the gravity of the theorems. The letter is written below.

8, February 1913
Trinity College, Cambridge

Dear Sir,

I was exceedingly interested by your letter and by the theorems which you state. You will however understand that, before I can judge properly of the value of what you have done, it is essential that I should see proofs of some of your assertions.

Your results seem to me to fall into roughly 3 classes:

(1) There are a number of results which are already known, or easily deducible from known theorems;

(2) There are results which, so far as I know, are new and interesting, but interesting rather from their curiosity and an apparent difficulty than their importance;

(3) There are results which appear to be new and important, but in which almost everything depends on the precise

Life of Ramanujan

rigour of the methods of proof which you have used.

I hope very much that you will send me as quickly as possible at any rate a few of your proofs, and follow this more at your leisure by a more detailed account of your work on primes and divergent series. It seems to me quite likely that you have done a good deal of work worth publication and if you can produce satisfactory demonstrations, I should be very glad to do what I can to secure it.

I have said nothing about some of your results-notably those about elliptic functions. I have not got them to refer to, as I handed them to another mathematician more expert than I in this special subject.

Hoping to hear from you again as soon as possible.

I am
Yours very truly,
G.H.Hardy

(Here a part of the letter is inserted excluding discussion about mathematical results. Another mathematician implies J.E.Littlewood)

In reply whatever Ramanujan wrote was the expression of his suppressed grief for a long time. He really got a person in G.H Hardy to whom he could express his sorrow and who could realize it.

The detail of the second letter to G.H. Hardy is written below.

27 February 1913
Madras Port Trust Office
Accounts Department

Dear Sir,

I am very much gratified on perusing your letter of the 8th February 1913. I was expecting a reply from you similar to the one which a Mathematics Professor at London wrote, asking me to study carefully Bromwhich's Infinite Series and not fall into the pitfalls of divergent series. I have found a friend in you who views my labours sympathetically. This is already some encouragement to me to proceed with my onward course. I find in many a place in your letter rigorous proofs are required and so on and you ask me to communicate the methods of proof. If I had given you my methods of proof I am sure you will follow the London professor. But as a fact I did not give him any proof but made some assertions as the following under my new theory. I told him that the sum of an infinite number of terms of the series:

$1+2+3+4+\ldots\ldots\ldots\ldots = -1/12$ under my theory. If I tell you this you will at once point out to me the lunatic asylum as my goal. I dilate on this simply to convince you that you will not be able to follow my methods of proof if I indicate the lines on which I proceed in a single letter. You may ask how you can accept results based upon wrong premises. What I tell you is this. Verify the results I give and if they agree with your results, got by treading on the groove in which the present day mathematicians move, you should at least grant that there may be some truths in my fundamental basis. So what I now want at this stage is for eminent professors like you to recognize that there is some worth in me. I am already a half starving man. To preserve my brains I want food and this is now my first consideration. Any sympathetic letter from you will be helpful to me here to get a scholarship either from the University or from the Government.

With respect to the mathematics portion of your letter it is the results that you class under 1st head and which you say are already known or are easily deducible from known theorems which encourage me now to proceed onward. For my results are verified to be true even though I may take my stand upon slender basis. I may now assure myself that my results and my methods of proof are as rigid as ether. Suppose I say that ether is rigid to one who does not know the ether hypothesis. He will simply laugh. The results I gave in my letter to you were only examples derived from substitution of particular values in some of the theorems I got. This time I give you some more general than the previous ones but still only particular cases of my theorems.

You may judge me hard that I am silent on the methods of proof. I have to reiterate that I may be misunderstood if I give in a short compass the lines on which I proceed. It is not on account of my unwillingness on my part but because I fear I shall not the able to explain everything in a letter. I do not mean that the methods should be buried with me. I shall have them published if my results are recognized by eminent men like you. You ask me to give you the expression I have got for the number of prime numbers within a given number. These are the expression that I have obtained for the number of primes less than a give [n] number.

With kind regards
Yours very truly
S.Ramanujan

(The Mathematical portion of the letter is not included which comprised about ten pages)

After receiving Hardy's letter, the higher education department of Madras began their activity to do something for Ramanujan. Mathematics professor B. Hanumantha Rao invited Narayana Iyer to the meeting of the Board of Studies in Mathematics to discuss about formulating an opportunity for Ramanujan. The Board came to a conclusion that they would request the Syndicate of the University to grant a scholarship of seventy five rupees per month which would be more than double of his salary in Port Trust of Madras. But a difficulty arose regarding the rule of the university as there was no provision to grant a scholarship to a student who could not cross the boundary of colleges. After many hot discussions, the rule was relaxed for Ramanujan as a special case and a scholarship of seventy five rupees per month was granted for two years. Thus Ramanujan became the first researcher of Madras University.

1.5 The Research student Ramanujan

April 12, 1913 became a very delightful day for Ramanujan, as he came to know that his scholarship was granted. He had no time to waste by working in the Port Trust office. By getting leave from his job, he would be able to devote the whole day to his mathematical research. Not only the scholarship but also some other good activities were waiting for him. He was set free to attend lectures of professors at the university as well as to use its library. This was the best arrangements for Ramanujan. If the boon had been granted a few years earlier, it would have been a blessing both for Ramanujan and the mathematics world. The age from 18 to 25 is very crucial for a mathematics researcher. We know how Ramanujan had spent this vital part of his life in a precarious condition. G.H Hardy's remark about these five years was, "During his five unfortunate years (1907-1912) Ramanujan's genius was misdirected, side tracked and to a certain extent distorted".

Ramanujan deserved that type of appreciation and recognition much before for the full bloom of his talent. He had no need to quit his service of Port Trust as he had been granted leave without pay for a period of two years. He remained free in his field of work but had to submit progress-report after every three months. Now, his income had become more than double, so he rented a spacious house in a lonely locality to accommodate his mother, grandmother and wife. The house was not far from Presidency College; it was within one and half miles.

Almost every evening he and Narayana Iyer remained in their mathematical preoccupation in this house. He used to go to sleep at night after requesting mother and grandmother to awake him in the dead of night, so that he might engage himself in mathematical work with a cool brain and in complete silence.

Janaki Devi, his wife had primary knowledge of Tamil language. She could write and read. Ramanujan had no interest in his mind regarding his wife's education. But he taught her the basic daily life science through experiments. One day he took two pans and a tube to teach her principle of siphon by experiment. This could be known from the reminiscences of Janaki Devi.

Her reminiscence gives us more information about Ramanujan. Ramanujan did not want to waste time by eating food, which also interrupts continuation of thought. So she or her mother-in-law fed Ramanujan south Indian food lump by lump while he remained busy with his mathematical research. Thus, neither pen nor deep thought could take rest. Except that whenever she awoke in the dead of night, she saw he had been pondering over mathematics and the scratching on the slate by slate pencil or stylus could be heard.

G.H.Hardy realized from the beginning that Ramanujan should be brought to England to acquaint him with the thorough and entire progress of up-to-date modern mathematics. Hardy also realized that he should have education according to university level syllabus which is possible only in England for a genius like him. Otherwise he would waste time and energy by rediscovering theorems discovered by other mathematicians many years ago.

G.H.Hardy urged several influential persons of Madras to send Ramanujan to England for his proper progress. Hardy was informed that it was not possible to send Ramanujan to England owing to Hindu religious superstitions. Though some superstitious ideas are not in vogue now it was awful about one hundred years ago. I am trying to give an idea. Crossing of the sea was a deadly sin. It was commensurate with eating beef or tearing off sacred thread. The consequence was disastrous as one had to suffer ostracism. The perpetrator had to suffer even after death, as none would attend his funeral.

Our father of the Nation Gandhiji faced religious obstacles before going to England to study law. He had to swear before departure. I hereby insert few lines from Gandiji's book 'The story of my experiments with truth': My mother, however, was still unwilling. She had begun making minute inquires. Someone had told her that young men got lost in England. Someone else had said that they took to meat; and yet another that they could not live there without liquor. 'How about all this?' she asked me. I said: Will you not trust me? I shall not lie to you. I swear that I shall not touch any of those things. If there were any such danger, would Joshiji let me go?'

'I can trust you' she said. 'But how can I trust you in a distant land? I am dazed and know not what to do. I will ask Becharji Swami'.

Becharji Swami was originally a Modh Bania, but had now become a Jain monk. He too was a family adviser like Joshiji. He came to my help, and said: 'I shall get the boy solemnly to take the three vows, and then he can be allowed to go'. He administered the oath and I vowed not to touch wine, woman and meat. This done, my mother gave her permission.

Some days Gandhiji had to go without food due to lack of vegetable meal while he was in England as a student.

In the meantime Ramanujan and Hardy exchanged two or

three letters. Hardy mentioned lacuna in his proof and urged Ramanujan to send proper proof of his theorems. Ramanujan proved his theorems by his own method and in brief so it was not possible for him to send proper proof for Hardy's comprehension. Moreover those proofs were done on slate and results were noted on paper.

Though there was religious obstacle Hardy did not give up on his decision of bringing Ramanujan to England. E.H.Neville, his colleague from his college would go to India to deliver twenty one lectures on differential geometry in Madras University. Hardy had a secret desire in his mind that E.H.Neville would convince Ramanujan to go to England and accompany him to Cambridge University.

E.H.Neville

Eric Harold Neville came to India at the end of December 1913 and arrived in Madras around New Year's Day in 1914. Ramanujan was introduced to him after the first lecture. They sat down thrice with Ramanujan's note books. Ramanujan got a trustworthy person in Neville. A friendship grew between the two contemporary mathematicians. Ramanujan requested Neville to carry the note books to England for verification in his spare time. He had protected the note books from boyhood and did not even send them to Hardy; now, he had no hesitation in Neville's carrying them. Neville had the credit of becoming trustworthy to Ramanujan within a few days.

Neville realized that the note books were priceless and beyond comprehension of ordinary mathematicians. Now, E.H.Neville proposed to Ramanujan for higher education and research in England. Ramanujan was always in a mood to leave no stone unturned for studying his favorite subject. He said, "I have no objection to go to England for mathematics but the expenditure is huge; who will bear it for me? Besides I have to sit for an examination there and my knowledge in English is very poor." Though he was excellent in his subject, he was afraid of appearing for an examination, as he was unsuccessful four times in the colleges of India. Another obstacle was his vegetarianism. Neville said, "Don't worry, money would not be a problem for your progress. You have not to sit for any examination. You would have to remain in your mathematical research only. Your vegetarianism would be respected". A few days back Komalatammal had dreamt a dream that his son was seated among some persons of white skin and goddess Namagiri advised her not to stand as an obstacle in the way of her son's progress. His mother, though an orthodox Hindu Brahmin woman, gave her consent for her son's going to England. He had got permission from his mother; nevertheless he had no peace of mind. He wanted direct consent from goddess Namagiri.

In this situation all his well-wishers were persuading him to be prepared to go to England. But Ramanujan was waiting for direct permission from goddess Namagiri.

At last one day Ramanujan set out for Namakkal, eighty miles west of Kumbakonam with Narayana Iyer. Narayana Iyer and Ramanujan slept on the temple ground for three nights. Nothing happened on the first and second nights. But on the third night Ramanujan rose from a dream and awoke Narayana Iyer at dead of night. He narrated to Iyer that he had received direction for foreign travel from goddess in a flash of brilliant light. A desire to go to England for mathematics and a profound devotion to goddess Namagiri created a situation that happened at dead of night. It had to happen.

All obstacles from Ramanujan's mind disappeared but Janaki's father i.e. his father-in-law was unwilling to let him go. He raised a question, "Is it not possible to study mathematics in India?" Mothers always remain anxious about their sons. Komalatammal was worried that his fussy health might not be keeping well in the biting cold of England, and the difficultly his son might face due to non-availability of Indian vegetarian food.

However, Ramanujan's going to England was not obstructed.

The proposed sea-voyage, according to some of Ramanujan's friends was just like transfer of dignity from Madras University to Cambridge University. Neville assured them that Cambridge University would try to firmly settle down Ramanujan to his due position in the mathematical world. It was not a fact of dignity or honor for Madras University or Cambridge University at all.

Hardy and Neville arranged to bring Ramanujan to England but the money could not be arranged. As he had no adequate qualification, the authority was unwilling to grant him so much scholarship for a foreign trip. He was granted a scholarship of seventy five rupees per month for two years for research in India but many doubtful questions arose regarding his ability for research in foreign institutions.

Ramanujan was lucky enough to have had several foreign champions. Richard Littlehailes, an Oxford educated professor of mathematics at Presidency College, asked Francis Dewsbury, registrar, to grant Ramanujan 250 pound per year scholarship and 100-pound for western dress and booking passage to England. He wrote, "Ramanujan is a man of most remarkable mathematical ability, amounting I might say to genius, whose light is metaphorically hidden under a bushel in Madras."

Then Lord Pentland was the governor of Madras. He ruled over forty million people at that time. He was a person with a view that a person should be given opportunity for his full development. He played a vital role when Ramanujan had obtained his first scholarship one year before. This time, he gladly recommended a higher scholarship for Ramanujan. Thus Ramanujan won the hurdle race of his life. Now, he might proceed.

Ramanujan received a second class ticket for his voyage from Binny & Co. on February 26. On March 11, Sir Francis wrote to the steamer company that Ramanujan might get vegetarian food on his voyage to England.

On March 14, Ramanujan accompanied his wife and mother to Egmore station of Madras to send them to Kumbakonam, his native town. His eyes filled with tears. It was uncertain when he would be able to meet them again.

One day in her mother-in-law's absence, while she was at the nearby temple, Janaki urged Ramanujan to take her with him to England. He refused to agree as he was influenced by Ramachandra Rao and others. Ramanujan told her that if she accompanied him he would not be able to concentrate on mathematics. Another reason was that she was so young and pretty that Englishmen would approach her with unwelcome intentions.

It would have been the best for himself and mathematics if he had taken Janaki to England. Cooked food by Janaki could have saved him from his untimely death.

At this juncture every person would be exhilarated by the opportunity but Ramanujan remained calm and quiet. Perhaps he followed the teachings of the Gita, the Hindu religious book in which God advises us to remain calm in every situation of life. One should do any work according to His wish without expecting result. He is the Supreme Authority, who gives us results.

Now, Ramanujan had to change himself into a European gentleman within a few days. His friends and well-wishers took the responsibility to transform him. Ramachandra Rao ordered a hair-cut in the European style. His kutumi, the long bunched-up knot of hair at the back of his head was cut off, though he was not pleased with his western hair-cut.

Richard Littlehailes was driving him around Madras town on his motorcycle to purchase European dress for Ramanujan according to his size.

Ramachandra Rao arranged to keep Ramanujan at the house of his friend who lived in the European- style. Here, Ramanujan got training at the dining table; how to use a knife and fork, which made him unhappy.

He was perplexed by western clothes purchased for him. He did not know how to wear them. Making the knot on his tie puzzled him. It became a scene of everyone's amusement present in the room. Later his mother could not recognize him when he sent a photo of him from England in suit and tie

A brand new ship S.S.Nevasa arrived in Madras harbor on March 15, 1914. Anyway once in1985, I came from Port Blair to Madras by ship and landed in Madras harbor; thus I got a chance to see the harbor from a ship. It is a big harbor.

Srinivasa Iyengar, the advocate general, arranged an official send off in Ramanujan's honor on the morning of his departure; Sir Francis Spring, Narayana Iyer, Professor Middlemast, Kasturirangar Iyengar, Publisher of the Hindu and many other respectable persons were present. Ramanujan was introduced to everyone present there. Here, Narayana Iyer who had worked closely with Ramanujan, exchanged his slate with Ramanujan's as a memorial. J.H. Stone, Director of Public Instruction of Madras, wished him success and told him that he had written to his friends in England who would take care of him.

When everyone was in a merry mood, tears rolled down

from Ramanujan's eyes, it was painful for him to leave friends, relatives and his country for an uncertain period.

At 10 A.M. the S.S Nevasa began to drift away from the dock. The remarkable day was March 17, 1914.

The S.S. Nevasa was quite big and its four decks were airy and spacious. Ramanujan had not travelled by a ship before, so his health was not keeping well due to seasickness. I travelled by ship several times and had the experience of problems regarding health due to rolling and pitching of the ship. A passenger would feel dizzy or giddy and nausea .Sometimes one is compelled to vomit due to the sway of the ship. Ramanujan got temporary relief when the ship anchored in Colombo harbor of Sri Lanka (then Ceylon) .Then the ship entered into Arabian Sea after crossing Cape Comorin. Now, Ramanujan began to enjoy his first trip to England as the sea was not rough. About two hundred passengers were travelling in the ship. Ramanujan conversed with some of them and got acquainted. But he was not a person like others who loved chatting away time. Most of the time he engaged himself in mathematics in his cabin.

The ship crossed the Arabian Sea and entered the Red Sea. Passengers sent letters to their homes whenever the ship anchored in ports like Aden, Port Said etc.

It was like an irony that a person, who was unsuccessful four times in Indian colleges, was going to Trinity College of Cambridge for research.

After crossing the Mediterranean Sea the ship anchored on Plymouth port of England .The voyage ended on April 14 when the ship arrived at the mouth of the Thames.

1.6 Ramanujan in England

The sky was clear, the sun was shining brightly, it was not so cold that day. Neville and his elder brother were present at the dock to receive Ramanujan. They drove him to the reception center in London. At first, all students from India were brought there. Ramanujan saw an astonishing sight from the moving car that people of white skin were engaged in manual labor. He had not seen any European performing manual work in India.

Ramanujan spent three or four days at the reception center, situated by the side of 21 Cromwell Road, in the South Kensington district of London. On April 18, Neville brought him to his house for his lodging. The house was situated on the bank of the river Cam in the suburb of Cambridge. Other students from India did not get familiar persons in England; most of the time they fell into an awkward position in an unknown society. Here, Ramanujan knew Neville. His wife Alice was so hospitable that Ramanujan felt no difficulty in a new environment.

He started his mathematical research with Hardy and Littlewood. He had begun to attend class lectures by Hardy, Littlewood and other professors. Every student in Cambridge was under a tutor who used to look after his student's progress. E.W.Barnes was engaged as tutor of Ramanujan. In his opinion Ramanujan was the most intelligent of all the top Trinity students during his tenure.

One day Professor Arthur Berry was solving problems of elliptic integrals on the blackboard in King's College. Ramanujan's face became bright as he was following the lecture. Professor Berry asked him whether he was able to follow his lecture or if he had any question to ask him. Ramanujan stood up and went straight to the blackboard. He wrote the results of the problem by chalk which was yet to be proved by Professor Berry. Later Professor Berry remarked: "Ramanujan must have reached those results by pure intuition."

The Talent of Ramanujan made him famous among the students within a few days. Most of the time he remained busy in his room. So he could not be seen outside. Sometimes he could be seen strolling alone in the spacious premises of Trinity College. He put on slippers while walking as he was not yet accustomed to wearing shoes.

Ramanujan came to the college hostel after dwelling for about two months with the Neville family. He shifted to hostel to devote more time in study. The hospitality that the Neville family rendered to him made a lasting impression on his mind.

When he was in India he had to send letters for Hardy's opinion. But in England he could meet Hardy regularly. Notebooks of Ramanujan embarrassed Hardy. But some of the theorems were rediscovered by Ramanujan like the ones discovered fifty or a hundred years ago by European mathematicians.

The notebooks contained more than three thousand five hundred theorems accumulated during ten years. The prudence of Hardy could easily understand that those theorems would need several generations of mathematicians to prove.

George Polya, a Hungarian mathematician took the notebooks from Hardy to prove the theorems of Ramanujan. He returned the notebooks and said that he would have to spend all his life if he tried to prove the theorems and would not be able to invent any theorem himself.

Later Professors G.N. Watson and B.M. Wilson could advance an insignificant amount after working for several years on Ramanujan's notebooks.

In 1977 Bruce C. Berndt, an American professor could advance to a small extent after working for several years. I have written a few lines above to give an idea about the vastness of Ramanujan's work.

One day Littlewood informed Hardy during a discussion that Ramanujan could be compared with the famous mathematician, Jacobi. Hardy advanced one step forward and told him that it was very difficult to think Ramanujan as equal to any other mathematician, but he could be compared with two famous mathematicians Euler and Jacobi. His profound intuition and invincible originality made him the greatest formalist of his time. If he was born one hundred years ago he would have invented most of the formulae and other mathematicians would get less scope to invent them.

Hardy was very proud of Ramanujan. Once he wrote: Ramanujan was my discovery. I did not invent him-like other great men, he invented himself- but I was the first really competent person who had the chance to see some of his work, and I can still remember with satisfaction that I could recognize at once what a treasure I had found.

P. C. Mahalanobis

Prasantha Chandra Mahalanobis, a twenty year old Bengali young man from Calcutta (now Kolkata) came to England to study in London in 1913. He came to Cambridge for sightseeing and felt an attraction for this town. He had met the head of King's College and was allowed to study there. Thus he became a student at King's College, Cambridge.

Once his mathematics professor asked him, "Have you met your astonishing compatriot Ramanujan?" Prasanta Chandra replied "I have not yet met him, but I have heard of him".

One night Prasantha Chandra came to meet Ramanujan. He saw him sitting close to the fire. He informed Prasantha Chandra that there were not enough blankets on the bed, so he used to sleep with his overcoat on, wrapped in a shawl close to the fire. Prasantha Chandra taught him how to use the blankets to get relief from the cold, as there were enough blankets on the bed tucked neatly under the mattress.

Ramanujan became famous among the people of India after his sudden departure to England. Newspapers took leading part to praise him after two months of his departure. Newspapers wrote: Mr. S.Ramanujan of Madras, whose work in Higher Mathematics has excited the wonder of Cambridge, is now in residence at Trinity. He will study mainly under the two fellows of the college- Mr. Hardy and Mr. Littlewood. They are going through the volume of work he has already done, and are making some surprising discoveries in it!

In England an intimate friendship was formed between the Tamil student and the Bengali student. Every Sunday morning, they used to walk a long distance. During that span of time, they engaged in discussion on various subjects of importance. Later Prasanta Chandra wrote that Ramanujan was not less competent in philosophy than in mathematics. Ramanujan sometimes said to his friends,"An equation for me has no meaning unless it expresses a thought of God".

One Sunday morning Prasanta Chandra was solving a problem in Ramanujan's room. Ramanujan was preparing his vegetarian food in the back kitchen. He was a Brahmin; moreover he was a Vaishnava. He was brought up in such an environment that he could not imagine eating non-vegetarian food. He did not eat vegetables like tomato, onion etc. perhaps due to their red color like blood. Even he never ate fried potato from the college kitchen if it was treated in lard.

Prasantha Chandra solved a problem through trial and error in a few minutes. Then he thought how much time Ramanujan would take to solve the problem. He said loudly, "Here's a problem for you." Ramanujan said while cooking "What problem? Tell me" Prasantha Chandra read it to him. Ramanujan solved the problem in an instant "Please take down the solution" said Ramanujan, still stirring vegetables. Prasantha Chandra was astonished by the intuition of Ramanujan.

Ramanujan (centre) with friends

Prasantha Chandra Mahalanobis returned to India after obtaining the degree. Later he became F.R.S. He was the founder of the Indian Statistical Institute. After independence of India he became a member of the Planning Commission.

Now, Ramanujan was a very happy person. Though he was away from home and motherland, he had no financial problem and dearth of food. He could remain in his mathematical researches without anxiety. He was getting help from Hardy and other mathematicians.

His papers were being published in journals and were praised. He wanted recognition of his research while in India but was deprived. In England he had got the fulfillment of his desire.

Ramanujan's life was just like a hurdle race. World war (1914-18) appeared before him as a great hurdle. England was one of the countries of Triple Entente, so its involvement was significant. Many countries even India were engulfed by the war. World War I did not spare Cambridge University from its blaze. Most of the young professors of the university were recruited in various departments of the military agency. Countless students from various colleges joined the military to fight for their country. Many professors and students were killed in the war. E.W. Barnes, Ramanujan's tutor, at the end of the war lamented: "Of my pupils at Cambridge at least one half, and practically all the best, have been killed or maimed for life; the work that I did [teaching mathematics over the years] has been for the most part wasted." Army camps and hospitals were erected in the university campus. Ramanujan wrote to his mother to lessen her anxiety about the war that was far away from his college campus but it was not so.

The war diminished his concentration on his study of mathematics. He was deprived of the guidance of many professors from different branches of mathematics. The war acted upon his health indirectly. Ramanujan was a Brahmin; over and above that he was a Vaishnava, so he had never eaten fish or meat from his childhood and never imagined doing so. His principal food was milk, butter, fruits etc. Except that he used to receive from home powdered rice and other ingredients for making South Indian food himself. As a result of the war, civil ship movement diminished and supply of ingredients for making South Indian food had been stopped. Supply of dairy products and fruits were also lessened. Those who were non-vegetarian felt no difficulty but Ramanujan was affected most. Scanty meals in a cold country gradually made Ramanujan weak and vulnerable to disease.

In the first year that is 1914 only one research paper of Ramanujan was published in the Quarterly Journal of

Mathematics under the title "Modular Equations and Approximations to Pi". This work was done by him while he was in India.

Next year his nine research papers were published in the journal and his talent was appreciated in various countries.

1.7 S. Ramanujan, B.A.

The research which Ramanujan was conducting in mathematics was beyond comprehension of a B.A. degree holder; nevertheless he himself had no degree. He had a dormant desire to have a degree as he was a failure four times in Indian colleges. In March 1916, he received his coveted B.A. degree for his long paper on highly composite numbers. He got the degree from one of the best universities of the world, the august Cambridge University.

Ramanujan could not make himself smart though he was among English students and professors for two years. On 18 March 1916, in convocation he was awarded the degree. A group photo of students in their academic robes was snapped to mark the occasion.

Ramanujan (centre) with , G.H. Hardy (extreme right), and others

In the photo he looks shabby in academic robes. The legs of his trouser are too short and the buttons of his coat were straining. But he was in the middle of the photo as he was the toast of all students.

Geniuses like Ramanujan are indifferent to worldly concerns. Sir Issac Newton, Albert Einstein were not regular regarding their meals and sleep. They used to forget to eat though food was served in front of them.

In India it was a regular scene that Ramanujan was absorbed in mathematics and Janaki Devi was feeding him South Indian food lump by lump. Otherwise he would forget to eat.

He never ate in the college dining hall due to his rigid vegetarianism. He himself cooked his vegetarian food. But availability of vegetarian food like milk, butter, fruits, vegetables etc. became scarce due to the war. He was irregular in his habits. Sometimes he kept himself engaged in mathematics for about thirty hours at a stretch and then slept for twenty hours. Due to his devotion to mathematics cooking was neglected and he starved or remained half-fed many of the days. He gradually became weak with the progress of a hidden disease.

Indian students in England were not less meritorious compared to the English students; nevertheless they were weak in health. English students used to play various games or did some physical exercise, but students from India were less interested in sports and games. So Indian students were prone to various diseases due to less immunity. Ramanujan had no interest in sports and games from his childhood; even he did not go to watch any game or match. He had no interest in any subject except mathematics. Though Hardy was a cricket player and a connoisseur spectator of all games, he had never encouraged Ramanujan in sports and games.

Sometimes during holidays he went to London with his friend Gyanesh Chandra Chatterjee, (who later became vice chancellor of Rajasthan University) where he visited the Zoo, the British Museum or enjoyed a drama with his friend.

Two amazing incidents happened in the presence of G.C. Chatterjee. Once three friends came to London for a few days on a holiday tour. They rented a flat on the top floor. Their meals were served on the ground floor. After having dinner, at first Ramanujan went to his room for doing mathematics. Within a few minutes he walked down the stairs. He informed G.C. Chatterjee that a cat was sitting on his bed and he could not drive it away. Then three friends went to chase away the cat. G.C. Chatterjee threw a bundle of newspaper towards the cat but it did not budge. At last they approached the mistress of the house for help. She became irritated being disturbed at night, she came and after knowing the cause of her disturbance, she said, "What class of people are you?" Going near the bed she caught hold the cat by its neck and drove it away. She left the room smiling and bidding "Good night".

Gyanesh Chatterjee was acquainted with Ila Rudra, a student at the teachers' college. They decided to marry shortly. Ramanujan invited his friend and friend's fiancée at a dinner in his room. He had never entered the kitchen while at home but in England he used to cook for himself. Now, he was boastful of his cooking. At dinner Mrilani Chattopadhyay, sister of Sarojini Naidu and a student of ethics at Newnham College of Cambridge, was also present as a companion.

He started with serving soup at the table. Highly delighted Ramanujan's second round serving of soup was accepted by all but the third offer was declined by the women guests. Their dining table discussion was in full swing and all were expecting Ramanujan's next cuisine but their host did not turn up. Three invitees sought for and realized Ramanujan had disappeared from the apartment. After waiting for an hour Gyanesh Chandra Chatterjee walked down the stairs and inquired after Ramanujan. He was told, Ramanujan had left the place in a taxi. Being embarrassed they waited until ten o'clock; by this time they had to return.

G.C.Chatterjee had to spend four dreadful days due to lack of information about Ramanujan. Suddenly he received a telegram from Oxford on fifth day requesting him to wire five pounds. Yes, it was telegram from S.Ramanujan. Next day, he arrived in Cambridge. G.C.Chatterjee asked him, "Why had you disappeared keeping me anxious?" Ramanujan replied "I was hurt and insulted when the ladies refused food I served. I fled as far away as I could with the money in my pocket and determined not to come back while they were in my room." This incident implies that Ramanujan was a sentimental prodigy.

Though Ramanujan was not keeping well, due to paucity of vegetable food and his irregular habits, he was not admitted in a hospital for his treatment.

One hundred years ago medical science was not as advanced as it is now. Though causes of some of the diseases were discovered but medicines were not invented. Moreover diagnostic system in those days was not scientific or up to the mark. Treatment of patients was performed on the basis of the doctors' conjecture.

In 1882, Robert Koch discovered that TB is caused due to infection by a particular microorganism, bacillus Mycobacterium tuberculosis.

Dr. W.C.Wingfield, medical superintendent of an English sanatorium observed that T.B. is caused by over work, overplay, over worry, undernourishment, lack of sunshine and fresh air or immoderate life style. Ramanujan led an intemperate life for his mathematics.

Britain was dependent on imports for her foodstuff. Movement of cargo ships diminished owing to sinking of ships by submarines. Poor Englishmen began to starve due to paucity of food and exorbitant price.

I had heard from my grandfather that India was also affected by World War I as India was then under British Rule. Dearth of every household commodities and escalating prices compelled poor people of India to suffer starvation. The arrival of ingredients that were sent by Ramanujan's family or friends to make South Indian food became uncertain owing to irregular cargo ship movement.

Though Robert Koch (1843-1910) discovered the bacillus in 1882, effective drug streptomycin was invented in 1950. T.B. patients were kept in open air treatment or natural treatment in open lodges exposed to the fresh air. Patients were given nutritious meals and prescribed bed rest or light exercise for natural recovery. Natural treatment was agonizing for patients, they were kept in rooms or halls without any doors or glass in the windows. We the people of a tropical country can easily understand the misery of T.B. patients at that time in a cold country.

Ramanujan was not keeping well from the beginning of 1917. In the month of May he was admitted to a nursing Hospital for five months. After that he was kept most of the time in various T.B. sanatoriums. At first he was admitted to Mendip Hills sanatorium for about three weeks in October 1917. Then he was shifted to Matlock House sanatorium in Derbyshire. Here, he was under the treatment of three doctors and expenditure was huge. It was unbearable for Ramanujan to put up with the open air treatment in the biting cold. He complained to G.H. Hardy through a letter:

I have been here a month and I have not been allowed fire even for a single day. I have been shivering from cold many a time and have not been to take my meals sometimes .In the beginning I was told that I could not possibly have any except the welcome fire I had for an hour or two when I entered this place. After a fortnight of stay they told me that they received a letter form you about one and promised me fire on those days in which I do some serious mathematical work. That day hasn't come yet and I am left in this dreadfully cold open room.

Ramanujan received no letter from his wife Janaki as she was unsuccessful in sending him any letter. Letters were intercepted by Komalatammal. Janaki tried several times to send letters to her husband concealed in food packets to be sent to Ramanujan. But every time, her mother-in-law threw away the letters whenever she found them in food packets thus Janaki also got no letter from her husband as those were confiscated by her mother-in-law. Father of Ramanujan was also annoyed by the unethical behavior of his wife.

Once Ramanujan wrote to his mother that if she would send Janaki to England, all difficulties regarding his meals would be solved. Without informing Janaki she refused the proposal. Komalatammal, mother of Ramanujan, made an awful blunder by refusing Ramanujan's proposal. The untimely death of our prodigy could have been prevented if his wife Janaki was with him.

Janaki had got an opportunity for going to her father's house at Rajendram on the occasion of her elder brother's marriage. After the ceremony, she accompanied the new couple to Karachi, the place of posting of his brother. From there she wrote a letter to Ramanujan asking for some money for purchase a sari for herself and a gift for her elder brother's bride.

Though Ramanujan sent money and a letter, he did not inform her of his illness and difficulties in the letter. He was offended as he had not received any information from Janaki and the refusal of his mother to send Janaki to England.

Ramanujan had interest in astrology. Sometimes he calculated auspicious time for any happy occasion for his friends or relatives. He sometimes said to his friends that he would expire before attaining the age of thirty five according to the horoscope calculated by himself. Though he was orthodox in his religion he had respect for all religions.

Ramanujan became weak and progress of his mathematical research was at a low ebb but he did not stop his research due to ill health.

Ramanujan was elected to the London Mathematical Society on December 6,1917. Two weeks later on December 18, Hardy and eleven other renowned mathematicians signed the certificate of a candidate for election that nominated Ramanujan to become a fellow of the Royal Society. E. W. Hobson and H. F. Baker had not responded to Ramanujan's letters from India in 1913 but now, they were among the eleven renowned mathematicians who nominated him to become a fellow of the Royal Society.

1.8 Ramanujan became F.R.S.

On January 24, 1918 the names of 104 candidates were read out at a meeting of the society. The name of Ramanujan was amongst them. It was obvious that only a few of them would be elected. Hardy, Littlewood and other mathematicians were sanguine that Ramanujan would be honored though he was only twenty nine years old. For an F.R.S. twenty nine was of premature age. The renowned mathematicians namely G.H.Hardy and J.E.Littlewood had been thirty-three years old when elected F.R.S. Except that the number of research papers of Ramanujan were not adequate but, of course, of greater importance.

At that time Sir Joseph John Thomson (1856-1940) who was popularly known as J.J.Thomson was the president of the Royal Society. He is famous for his work on cathode rays, which led to the discovery of the electron in 1897. He continued to study the conduction of electricity through gases and for this work he was awarded the Nobel Prize for physics in 1906.

J.J. Thomson wanted to know from G.H.Hardy the circumstances surrounding Ramanujan's candidature. Hardy admitted that Ramanujan was beyond comparison among all other mathematical candidates; his work carried greater importance. His name could have been nominated a few years later but Hardy did not want to lose time owing to Ramanujan's deteriorating health.

In the meantime Ramanujan created a horrible incident devoid of sense. He leaped before an incoming train in a nearby station. No sooner had he jumped on the rail from the platform than the driver applied the brakes and succeeded in stopping the train just before him. It was possible for the driver and the guard to stop the train quickly as it was in slow motion, and they could see him leaping in front of the train. Ramanujan got badly injured and bleeding started from his affected knees and other parts of his body.

He was arrested by the police and driven to Scotland Yard, headquarters of the London police. In response to a call, Hardy reached there and tried to convince the police that they had done a great mistake by arresting Ramanujan, a Fellow of the Royal Society. An F.R.S. could not be arrested.

Till then Ramanujan was not elected F.R.S. and the police were not fooled. They could not rely on Hardy. After investigation they found that Ramanujan was a genius mathematician and they did not want to spoil his life. The police then decided to set him free.

Ramanujan came to England for mathematics but he was unable to adjust with the people and their culture. He had no intimate friend and even Hardy could not become his bosom friend due to the difference of culture and country. He wished Janaki would stay with him but he had not got even a letter from her. His mother could not understand the difficulties faced by her son in an alien country. He was alone in a foreign country with fragile health and without his favorite South Indian food. Moreover he was living in an awful World War I environment. College Campuses were full of army camps. Many professors and students joined the army. The number of students became thin. News of casualties perhaps had been troubling his mind. In these circumstances he allowed his body to fall before an approaching train.

Late in February, Ramanujan received the good news that he had been elected to the Cambridge Philosophical Society. On February 28, 1918 at an important meeting, the Royal Society elected fifteen Fellows out of 104 candidates and Ramanujan was one of them and the youngest. He came to know about this coveted honor by a telegram sent by Hardy from London while he was at Matlock. At first he had been confused by the telegram and repeated reading made it clear to him that it was not the Philosophical Society but the Royal Society that elected him a fellow, which Hardy wanted to inform. From May 2, 1918 he would become S. Ramanujan, F.R.S. Being an F.R.S. meant scientific distinction. Young scientists coveted for it and senior scientists lamented for lack of it.

Royal Society is the oldest scientific society in Britain. It was founded in 1660 and its early members included Robert Hooke, Christopher Wren, Isaac Newton, and Edmond Halley. It provided an impetus to scientific thoughts and developments in England, and its achievements became internationally famous.

Nobel Prizes have been awarded since 1901 for physics, chemistry, physiology or medicine, literature and peace. A sixth prize has been awarded in economic sciences since 1969. Till now there is no provision for awarding Nobel Prize for mathematics. Mathematicians like Ramanujan was worthy of being awarded the Nobel Prize.

Ramanujan became very grateful to Hardy, after being F.R.S., he wrote to him: My words are not adequate to express my thanks to you. I did not even dream of the possibility of my election.

His success came as sensational news to India. Everyone was delighted. On March 22, the Madras members of the Indian Mathematical Society conveyed thanks to Hardy.

Ramanujan became the first Indian F.R.S. in mathematics in 1918. In 1841 Ardaseer Cursetjee who was a marine engineer was the first Indian F.R.S. So Ramanujan became second Indian F.R.S. but the first Indian F.R.S. in Mathematics.

Ardaseer Cursetjee (1808-77) was the first modern engineer of India. He came from a family which had a long history of service to the British in the field of shipbuilding. His forefathers migrated from Surat to Bombay. The use of steam engine in navigation was introduced from the early 19th century. He became more interested in steam engine than in shipbuilding though shipbuilding was his family business.

Cursetjee installed a one Horse Power engine on his premises to pump water from a well to feed a small fountain. This engine built by him was the first engine built in India. He had a small foundry at his residence. He fabricated many tanks (up to 5000 gallons capacity) for different ships. He installed a 10-HP marine engine in a steamer named Indus in 1833. The engine was obtained from England. Indus was the 2nd steamer built in Bombay, after the Hugh Lindsay four years before. The Government of Bombay purchased his vessel and he was made assistant builder at Mazagaon.

His next engineering marvel was the installation of gas lighting in his bungalow at Mazagaon. The governor visited his residence and presented him with a dress of honor.

Cursetjee was requested by Elphinstone Institution for part time services to give practical lessons in mechanical and chemical science as they had no competent teacher to give practical lessons. Royal Asiatic Society of England elected him a non-resident member. He intended to stay in England for a year for upgrading his knowledge of marine steam engines. His departure was delayed due to his sudden illness and he missed the passage on a government ship. He left next year paying rupees one thousand for the passage.

Cursetjee maintained his life according to Parsi tradition. He ate food cooked only by Parsis so he took his servants along with him to England. He never shared a table with a non-Parsi person. Donning the traditional Parsi cap even in England was a must for him. He appealed to the liberality of the court of directors for an allowance to meet his expenses in England. He was sanctioned a daily allowance of one pound for a year over and above his Bombay salary of rupees seventy nine per month.

Ardaseer Cursetjee's sojourn in England was full of achievements. He became a member of the Society of Arts and Science. He was made a member of the mechanical section of the British Association for the Advancement of Science. He became an associate of the Institution of Civil Engineers. Cursetjee got an appointment as the chief engineer and inspector of machinery in the company's steam factory and foundry at Bombay. His salary enhanced from Rs. 79 per month to Rs.600 per month. While in England in 1841, he was nominated by influential persons for the fellowship of the Royal Society.

Cursetjee took his new charge on 1st April 1841, after returning from England. He became the first Indian to be placed over Europeans. Among his staff he had one chief assistant, four European foremen, one hundred European engineers and boiler makers and about two hundred Indian skilled mechanics. A controversy was raised on appointing an Indian to direct Europeans in an establishment as the Bombay Steam Factory. He made a success of his career and proved his competency.

Cursetjee visited America and arranged to bring woodcutting machines to Bombay. Sewing machines and electroplating were introduced by him at Bombay. He became the first Indian fellow of the Royal Society on 27 May 1841. Adraseer Cursetjee did not get due honor in independent India.

A.S. Ramalingam, South Indian engineer who was working in England, became acquainted with Ramanujan just after his arrival from India, at the Cromwell Road reception Centre. They had not met again since then but news of Ramanujan's achievement made him proud as a South Indian. He visited him at Matlock Sanatorium on Sunday, June 16, to convey him congratulation and to make enquiries about his health.

A.S. Ramalingam

He stayed there for three days with Ramanujan. Their discussion covered topics like World War I, condition in India and so on. He realized in three days that Ramanujan was extremely fastidious about his food and it affected his health. He even did not drink Ovaltine or any other health drink lest they should contain any animal product. At the time of departure he requested Ramanujan to give up his fastidiousness for his health and mathematics. "Ramanujan was killing himself by slow starvation. He was thinking of his vegetarianism even at the expense of his health and life" A.S. Ramalingam wrote to Hardy.

In this connection I intend to write a few lines about Swami Vivekananda. He was a religious reformer. He preached harmony of religions and divinity of mankind. He said that each soul is potentially divine, omnipotent and omniscient. The aim of life is to realize the same through selfless service to humanity and assimilation of wisdom through introspection. According to him "religion is the manifestation of the divinity already in man".

After being a saint he travelled throughout India mostly on foot. He was shocked by the precarious condition of the poor who were then a majority of the Indian population. They had no food to eat and clothes to wear; moreover superstition in the name of religion worsened their condition. During his travels, he surmounted all superstitious beliefs, ideas and practices. He set out from Calcutta (now Kolkata) as a penniless saint. Sometimes he had to go without food for two or three days and survived on food offered by people of high and low castes. Swami Vivekananda consumed food offered by scavengers and enjoyed tobacco from their hookah.

In 1893 he reached America to participate at the World's Parliament of Religions, Chicago. There he lost his credentials and the small amount of money which he had brought with him. He consumed food to survive. This food was offered by people but he did not refuse meat and fish.

During his short span of life he taught us to rise above all barriers. He delivered his first lecture at the world's Parliament of Religions on 11 September, 1893. I am quoting a few lines of the lecture: As the different streams having their sources in different places all mingle their water in the sea, so, O Lord, the different paths which men take through different tendencies, various though they appear, crooked or straight all lead to Thee. Thus we learn religion is a subtle way of life but does not depend on food.

At last World War I came to a close on November 11, 1918. I have been writing this book during 2014 which is the centenary year of the World War I. When Britain declared war on Germany on August 4, 1914, she had only about 150000 combat ready troops and France had an army of 1290000. In comparison Germany had an army of 1900000. The only alternative to overcome the crisis was to rely on the Indian Army.

India recruited 1440437 men and sent 1381050 of them overseas to fight for the British Empire between 1914 and 1918. Indian soldiers fought tooth and nail saving Britain and her allies from a humiliating defeat. Indian troops played a decisive role in World War I and World War II but India's contribution was misinterpreted.

Dominick Dendooven, celebrated author of France expressed his view. "This is the only time your history meets ours. And it's remarkable it happened 100 years ago when transport and communication were not developed and racial bias existed. Without India, the war might have ended in a resounding German victory. Neither we nor you should forget that."

More than 140000 Indian soldiers fought to defend the French soil and thousands sacrificed their lives while doing so.

G.H Hardy pondered to send Ramanujan to India, at least for a few months for his recovery. On November 26, G.H. Hardy wrote to Francis Dewsbury in Madras regarding temporary return of Ramanujan to India for recovery of his health.

After the end of war Ramanujan was brought to another nursing home near London. It was called Colinette House, situated on the bank of the Thames. It was spacious and decorated. Now, Colinette became approachable for Hardy so he could visit Ramanujan more easily in a cab, frequently.

Pitiable number thirteen (13) though it is innocent but we the people have given it a designation 'unlucky thirteen'. My marriage was celebrated on March thirteen. Due to early arrival of northwester a strong wind followed by a hailstorm made the weather of afternoon very bad but the night was cool and the marriage party was fine. No untoward incident happened on that day and no crack in married life appeared even after twenty five years of marriage.

Though G.H. Hardy was an atheist, but he was fastidious about numbers. One day Hardy came to visit Ramanujan by a taxi and observed its number, 1729. Conveying a hello to Ramanujan he expressed his bafflement about the taxi number and expected a reply that it was not a bad omen.

Ramanujan appeased Hardy while lying on bed. "No Hardy, it is a very interesting number. It is the smallest number expressible as the sum of two cubes in two different ways".

For example,

$$1729 = 12^3 + 1^3$$
$$1729 = 10^3 + 9^3$$

We can get many numbers that are the sun of one pair of cubes.

For example,

$$35 = 2^3 + 3^3 = 8 + 27$$
$$91 = 3^3 + 4^3 = 27 + 64$$

But 35, 91 etc. numbers could not be expressed as a sum of another two cubes until we reach 1729. 1729 is the smallest number expressible as the sum of two cubes in two different ways. The number 1729 has a title; it is called Ramanujan's Number.

While in India Ramanujan researched with various numbers and recorded it in his notebook, it may be said he had a friendship with numbers. For that endeavor he could answer Hardy instantly.

Ramanujan was honored by his own college on October 10, 1918. Trinity College elected him a fellow; he became the first person from India to achieve this honor from that college. For this election he was awarded a fellowship of 250 pounds per year for six years. He received good news from India that Port Trust had extended his leave of absence and as he had become an F.R.S., the University of Madras had granted him a 250 pound per year fellowship for six years. Thus Ramanujan would get two fellowships 250-pound per year each for six years.

He became reminiscent after getting two fellowships. He had to spend his childhood and youth in a precarious financial condition; had not got a job to survive and continued his mathematical research under the patronage of R. Ramachandra Rao. Now, he had much money but he repented his diminished activity over the past two years. He wrote a letter to Dewsbury on January 11, 1919, which implies his mental state.

Sir,

I beg to acknowledge the receipt of your letter of 9th December 1918, and gratefully accept the very generous help which the university offers me. I feel, however that after my return to India, which I expect to happen as soon as arrangements can be made, the total amount of money to which I shall be entitled will be much more than I shall require. I should hope that, after my expenses in England have been paid, £50 a year will he paid to my parents and that the surplus, after my necessary expenses are met, should be used for some educational purpose, such in particular as the reduction of school-fees for poor boys and orphans and provision of books in schools. No doubt it will be possible to make an arrangement about this after my return.

I feel very sorry that, as I have not been well, I have not been able to do so much mathematics during the last two years as before. I hope that I shall soon be able to do more and will certainly do my best to deserve the help that had been given me.

> I beg to remain, Sir,
> Your most obedient servant,
> S.Ramanujan

For returning to India he needed a passport. A passport photo was snapped on February 24, 1919. As he was ill, he had become a thinner man. He sat for the photo, wearing his shirt and coat but his dress had become two sizes too big. His shirt, buttoned to the top, was loose around the neck. He had been bulky when he came from India. Now, layers of fat had disappeared from his body. Later when Hardy first saw the photo in 1937 he wrote, 'He looks rather ill, but he looks all over the genius he was.'

The Government of India issued a postal stamp in 1962 on the 75th birth anniversary of Ramanujan. It was a replica of that photo, another postal stamp was issued in India in 2011, on the occasion of his 125th birth day. The other stamp was issued in 2010.

His passport had been made ready. His name, age, profession etc. were recorded in it. Profession: 'Research student' was inscribed in the passport.

He embarked on the ship S.S. Nagaya for returning to India on 13th March, 1919. Among other things, he carried with him a dozen of books, a leather trunk filled with papers and a box of raisins for his younger brothers. While Ramanujan had been returning to India by ship his two short notes appeared in the proceedings of the London Mathematical Society on the same day March 13, 1919.

He revealed new congruence properties of the partition function and a new link between the first and second Rogers - Ramanujan identities.

All newspapers of Madras printed the news of his return from England. His emergence into fame was one of the topics of discussion among all his old school friends and professors. Madras University had already decided to offer him the post of a professor.

Before his arrival in Bombay, the Indian Mathematical Society met there for its conference. All speakers praised Ramanujan for his elevation to the Royal Society from his very humble origin. The Journal of the society printed the news of his return on April 1, 1919.

1.9 Ramanujan's Return to India

He disembarked from S.S. Nagoya in Bombay on March 27, 1919. His mother Komalatammal and brother Lakshmi Narasimhan were waiting at the dock to receive him. They set out from Madras on the twenty first. Just after alighting from ship he asked his mother, "Where is she?" 'She' implies Janaki. Mother was irritated by this question and replied, "Why annoy me over Janaki?" Domestic conflict had affected him in England. He realized that his arrival would not give him any contentment.

About one year ago Janaki went to her father's house on the occasion of the auspicious marriage ceremony of her brother. She had not returned and none of her father-in-law's house tried to contact her.

Lakshmi Narasimhan did not know where Janaki was staying. She might be in Rajendram or with her sister in Madras, expecting Ramanujan's arrival. So he posted two letters to Janaki asking her to meet Ramanujan in Madras as he wished. Janaki had already come to know about Ramanujan's return from the newspapers. She requested her brother to take her to Madras. R.Srinivasa Iyengar, her brother, had forbidden her not to rejoin the family as her mother-in-law hated her. After receiving Lakshmi's letter Janaki decided to go to Madras with her brother.

As Ramanujan was sick, after a few days rest in Bombay, he, his mother and brother boarded the Bombay Mail to Madras. They got off the train at Madras Central station on April 2. R.Ramachandra Rao and his old student K.S.Viswanatha Sastri, who had seen him off from Madras five years ago and whom Ramanujan had tutored in math, were present at the platform to receive him. Seeing the appearance of Ramanujan R. Ramachandra Rao was stricken with grief. He realized that Ramanujan would not live long.

Absence of Janaki at the station disappointed Ramanujan. He was helped to board a 'Jutka' a two wheeled, horse-drawn vehicle to take him to a bungalow of a lawyer situated on Edward Elliots road, at a distance of three miles from the station. K.S. Viswanatha Sastri followed on behind on a bicycle. After reaching the bungalow he began to eat South Indian food, yogurt and sambhar. He told Viswanatha "If I had this in England, I would not have got sick".

Many important and respectable persons began to visit him and offered him houses to dwell and money for treatment. Sir Franics Spring arranged to publish his full biography in a newspaper which appeared on April 6. He was offered a university professorship, which he said he would accept after recovery. His doctor M.C. Nanjunda Rao recommended shifting of Ramanujan to a lonely place as he disliked the throngs of visitors. So he was moved to a place called Venkata Vilas on Luz Church Road.

On April 6, Janaki and her brother reached Venkata Vilas and about a week later Ramanujan's father, grandmother and younger brother reached there from Kumbakonam to stay with Ramanujan.

When Ramanujan sailed to England, Janaki was only thirteen years old. Now, she had become eighteen years. Due to restriction imposed by her mother-in-law she could not even talk with her husband but now, they could talk and stay as a married couple. During discussion they could understand how Komalatammal had intercepted their letters. But she had become very anxious about her husband's health. Ramanujan had become thin, weak and lighter in complexion. He often coughed up phlegm indicating his severe illness.

Friends of Ramanujan and his wife Janaki noticed a change in his personality. He was no longer cheerful and affectionate; on the contrary he became depressed and gloomy. His devotion to god and temple diminished. He drank coffee now, to which he was indifferent before going to England. He had become peevish, impatient and petulant. He was not as obedient as before to his mother. Often he disobeyed his mother's suggestion. Family feud kept Ramanujan disturbed during his last year of life. Sometimes the quarrels were between Janaki and her mother-in-law, sometimes between Janaki and Ramanujan's grandmother. The bone of contention between Ramanujan and his mother was the distribution of money to poor students from the part of his fellowship. Komalatammal's intention was to retain all the money for the family. Janaki was also not indifferent to money or ornaments.

Though Madras city is situated on the bank of the Bay of Bengal it is not pleasant in summer. Daytime temperature rises to forty or forty two degree centigrade. Doctors advised Ramanujan to shift to a cooler place like Coimbatore, located in a hilly area. But Komalatammal wished Ramanujan to stay at Kodumudi, a little town famous for its Magudeswara temple. Ramanujan stayed there for two months. Doctor C.F.Fearnside, Divisional Medial Officer, visited him every Sunday. Janaki had already taken the responsibility of nursing her husband day and night without any shortcoming. She gave him medicine at the accurate time according to the prescription.

Ramanujan often told her, "If only you had come with me to England, perhaps I would not have fallen ill".

When the temperature dropped after summer, on September 3, Ramanujan left Kodumudi with his family and arrived at Kumbakonam next day. Komalatammal already had arranged spacious house for accommodation of the members of family and relatives as well as for her son's proper care. That arrangement had to be done as their family home was not spacious.

Doctor P.S. Chandrasekhara Iyer, a professor of hygiene and physiology at Madras Medical College and a renowned tuberculosis specialist, was selected for Ramanujan's treatment.

One day Dr. P.S. Chandrasekhara accompanied one of Ramanujan's friends, Sarangapani to the house where Ramanujan lay sick. Dr Chandrasekhara examined him for more than an hour and came to the conclusion that Ramanujan had been infected by tuberculosis. One day Ramanujan told his friend Sarangapani, "I have a friend who loves me more than all of you who does not want to leave me at all, it's this tuberculosis fever".

Dr. Chandrasekhara wanted that Ramanujan should stay in Madras town where it would be convenient for him to treat him. But he had no intention of leaving his native town Kumbakonam. Ramanujan had been admitted to several nursing homes in England and was treated by renowned doctors. In India, for nine months, he was treated by specialist doctors and changed places or towns to get relief but all went in vain. Deterioration of health continued unabated. Now, he had given up the will to live which is very bad for a patient's recovery.

Dr. P.S.Chandrasekhara

Ramanujan could not deny repeated insistence, so he gave consent for going to Madras in winter. Perhaps in the beginning of 1920 he arrived in Madras with his mother, wife and brother-in-law Srinivasa. Here they started to live in a house named Kudsia, then moved to another house named Crynant. Finally they moved to another house named Gometra.

Ramanujan had not written any letter to Hardy after he arrived in India. On January 12, 1920 he wrote to Hardy informing him about his last discovery of the mock-theta functions. A part of the letter:

I am extremely sorry for not writing you a single letter up to now..... I discovered very interesting functions recently which I call "Mock" theta functions. Unlike the "False" theta functions (studied partially by Prof. Rogers in his interesting paper) they enter into mathematics as beautifully as the ordinary theta functions. I am sending you with this letter some examples.

None would imagine that a person who was in sick bed, it would be better to say deathbed, could engage in research or discover a profound mathematical theory. It was possible only for Ramanujan. He defeated everything by his determination and perseverance. Many mathematicians expressed the opinion that his discovery of mock theta function on his deathbed was enough to make him immortal in the mathematical world.

Eminent mathematician G.N. Watson paid homage to the genius of Ramajujan for his ultimate discovery by the following sentences. He wrote : Ramanujan's discovery of the mock-theta functions makes it obvious that his skill and ingenuity did not desert him at the oncoming of his untimely end. As much as any of his earlier work, the mock theta functions are an achievement sufficient to cause his name to be held in lasting remembrance.

P.V. Seshu Iyer who was professor of Ramanujan in Kumbakonam Government College wrote about his mock-theta function: There are no papers and researches of his more valued nor more intuitive than those which he thought out during these fateful days. His physical body was failing no doubt but his intellectual vision grew proportionately keener and brighter.

Ramanujan was in his deathbed; in these circumstances we would not be able to conjecture the mental condition of his mother, wife and other relatives. Leaving no stone unturned Komalatammal met with G.V. Narayanaswamy Iyer, a high school teacher and Ramanujan's friend. He was also a famous astrologer. After studying the horoscope of Ramanujan he conveyed the opinion that the chart indicated either of a man who would die at the height of his fame, or one who, if he did live long, would remain obscure. Thus he was unable to give any solace to the sorrowful mother.

We could know about his last few days from the reminiscence of Janaki. Ramanujan became skin and bones before his death. He complained of severe pain in stomach and legs.

She would heat water in brass vessel and apply warmth and moisture of a towel to lessen pain or discomfort. His pathetic condition could not prevent him from working. He used to ask for slate from Janaki, remained working lying in bed, head propped up on pillows.

He transferred his results in papers. Janaki had the duty to keep the mathematics papers in the big leather trunk which Ramanujan had brought from England. He remained absorbed in mathematics and would not talk to anyone who would come to visit him. He was scribbling on his slate even when only four days were left before his death.

Ramanujan became unconscious at dawn on April 26, 1920. Janaki had not budged from her husband's bed and was feeding him sips of dilute milk. After about two hours he breathed his last. Then he was thirty two years four months four days old. We the Indians lost a rare genius at an early age before blooming fully.

Janaki Devi

Examining the case history of his disease, the doctors in later years diagnosed that Ramanujan died of Hepatic Amoebiasis which affected his liver. Bloodstrain was absent in his cough and spittle so Dr. Kincaid, principal doctor of Matlock House, doubted tuberculosis of Ramanujan. Some doctors gave the opinion that he had got blood poisoning or deficiency of vitamin B_{12}.

Funeral of Ramanujan was arranged at noon, none of his orthodox relatives came forward to the cremation ground near Chetput. He had perpetrated a sin by crossing the sea and a trip to Rameswaram temple, which is situated in the sea and connected by rail and road, for the purification planned by his mother was postponed due to his weak health.

R.Ramachandra Rao arranged the funeral with the help of Rajagopalachari. Rajagopalachari was his son-in-law as well as Ramanujan's boyhood friend. His mortal body was put to the flames at 01.00pm. His mortal remains began to dissolve in nature slowly.

Ramanujan left behind him a vast treasure of mathematics for the next generation of humankind. Perhaps he could not assess during his lifetime the value of the treasure which he left behind for us.

It is said that mathematics is the mother of science. Now,

theorems and equations of Ramanujan are being applied in various branches of science and technology. Computer science, polymer chemistry even cancer research depend on Ramanujan's work. In his time flight into space was a dream of humankind, his mathematics is being applied in space technology and a dream has been converted into reality. Do engineers want to make better blast furnace for iron extraction? They have Ramanujan's formula. Scientists of the next generation are astonished by the work of Ramanujan. His formulae are applicable to the subjects which were unknown in Ramanujan's time.

The French mathematician Laplace, who had carried out research on gravity, conceived the notion of 'black hole' but he did not use the term 'black hole'. A black hole is a celestial object whose mass concentration generates a gravitational attraction so strong that not even light can escape from its surface. So a black hole is the ultimate in darkness. The radius of our earth must be less than 8 millimeters if it is to become a black hole. For the sun its radius will have to be less than 3 kilometers to become a black hole. All celestial objects do not end up into black hole. There is a limit of the mass. Indian-born Nobel laureate in physics, Subrahmanyan Chandrasekhar found that there is a limit to the mass that can be held in equilibrium through the degeneracy pressure arising from the close packing of electrons. His limit, known as the Chandrasekhar limit, is about 1.4 solar masses. If a stat is more massive than this limit, its interior does not become degenerate, and the star will begin to shrink under its force of gravity.

Scientists had a notion that a black hole is the ultimate in darkness since light can't come out of the black hole. In 1974 Professor Stephen Hawking of the University of Cambridge after a series of calculations concluded that a black hole can radiate weak energy. This process is known as Hawking radiation.

Ken ono of the University of Atlanta has found a formula of Ramanujan that can disclose black hole mystery and corroborates

Hawking radiation. Ramanujan was advanced by a century or so in the field of mathematics and science. Only great mathematicians can judge his merit.

A dark young man, barely twenty years old, set sail from Bombay (Mumbai) on July 31, 1930 for studying in Cambridge. During his voyage he spent most of the time sitting on deck reading books and doing calculations. When he reached London he had worked out a result in astrophysics- the fate of stars.

The young man was Subrahmanyan Chandrasekhar. We celebrated his birth centenary in the year 2010. He was born in Lahore on 19.10.1910 where his father had been serving as a railway auditor. Their family had moved to Madras (now Chennai) by the time he was eight. He had got inspiration from his uncle C.V.Raman, who won the Nobel prize in 1930 for his discovery of Raman Effect.

The Raman Effect deals in scattering of light by molecules of a medium when they are excited to vibrational energy levels.

After completing his study in Madras he moved to England for higher education. He secured a scholarship to study at Trinity College in Cambridge where Ramanujan was a student a few years back.

Subrahmanyan Chandrasekhar

His discovery that he worked out during his voyage to England was not accepted for decades by other scientists. I got an opportunity to attend a lecture by British astrophysicist Sir Roger Penrose while he visited Kolkata in January 2011. Sir Roger Penrose said, "He showed that a white dwarf star would not be able to sustain itself from unstoppable collapse if its mass were greater than 1.4 times the mass of the sun" (known as the Chandrasekhar limit) .This led to the understanding that black holes could form in astrophysical processes.

Renowned scientists led by Einstein, Bhor, Heisenberg, Pawli and others had turned down the Newtonian world upside down. Indian scientists like Meghnad Saha and Satyendra Nath Bose were parts of this quantum revolution. Chandrasekhar spent a year in their company. Chandrasekhar received the Nobel Prize for physics on19-10-1983.As he received the Nobel Prize on his birth day; he jokingly called it a birthday gift. Chandrasekhar discovered Chandrasekhar limit but his knowledge was not limited to astrophysics only. He was a person with profound knowledge in many subjects. He loved literature, music and art .Here is a quotation from him:

To me, science is the search for truth, a definitive repetitive pattern in creation ....Mozart looked for it in music, Monet saw it in a haystack he painted, Michaelangelo found it in stone.....Perhaps a scientist seeks it through a mathematical equation. But we can all see only a part of the pattern at a time. It is given only to a few to see the whole.

We know that Ramanujan tried to find truth in his equations: "An equation for me has no meaning unless it expresses a thought of God"

Chandrasekhar was about nine years old at the time of Ramanujan's death in1920. His mother informed him of Ramanujan's death from the newspaper. He had no idea about Ramanujan or his mathematics at that tender age. In later years he got more information about Ramanujan in the Trinity College and in his meeting with Hardy.

After about seventy years of Ramanujan's death Chandrasekhar told an American audience, "I can still recall the gladness I felt at the assurance that one brought up under circumstances similar to my own could have achieved what I could not grasp.

The fact that Ramanujan's early years were spent in a scientifically sterile atmosphere, that his life in India was not without hardships, that under circumstances that appeared to most Indians as nothing short of miraculous, he had gone to Cambridge, supported by eminent mathematicians, and had returned to India with every assurance that he would be considered, in time, as one of the most original mathematicians of the century-these facts were enough, more than enough, for aspiring young Indian students to break their bonds of intellectual confinement and perhaps soar the way that Ramanujan had."

E.H. Neville was the first British Mathematician from Trinity College who met Ramanujan in January 1914, in Madras. Notebooks of Ramanujan astonished him and he persuaded Ramanujan to go to England according to Hardy's instruction. In 1941, Neville wrote a lecture on Ramanujan for radio broadcasting. He could not deliver in full the lecture during the scheduled time. The unaired portion of the lecture is appended below.

Ramanujan's career, just because he was a mathematician, is of unique importance in the development of relations between India and England. India has produced great scientists, but Bose and Raman were educated outside India, and no one can say how much of their inspiration was derived from the great laboratories in which their formative years were spent and from the famous men who taught them. India has produced great poets and philosophers, but there is a subtle tinge of patronage in all commendation of alien literature. Only in mathematics are the standards unassailable, and therefore of all Indians, Ramanujan was the first whom the English knew to be innately the equal of their greatest men. The mortal blow to the assumption, so prevalent in the western world, that white is intrinsically superior to black, the offensive assumption that has survived countless humanitarian arguments and political appeals and poisoned countless approaches to collaboration between England and India, was struck by the hand of Srinivasa Ramanujan.

1.10 Ramanujan's relatives and mentors after his death

Janaki Devi remained alive for seventy four years after her husband's death. If Ramanujan had been alive she would not have been deprived of wealth and dignity. It was a pity that she had to struggle with her misfortune throughout the rest of her life. In those days life of Brahmin widows was very miserable. They had to remain without a drop of water and food on the eleventh day of the lunar fortnight. My grandmother's mother became a widow when she was twenty one years old in 1921. Her husband died of T.B. as there was no treatment for T.B. in those days. I saw her in her old age when I was a child. She used to eat once in a day at noon. I can still remember her restlessness for water on ritual days in the summer.

After the cremation of Ramanujan, Janaki Devi returned to Rajendram with her mother as she could expect no help from Ramanujan's family. Then she lived with her brother in Bombay for six years. Her brother was an income tax officer there. She had no educational qualification, no skills, so she began to learn embroidery and tailoring in Bombay. She returned to Madras after the University of Madras awarded her a pension of twenty rupees per month. The pension was arranged when she gave up her rights to Ramanujan's papers.

Staying for a few days with her sister in Triplicane, she shifted to a house on Hanumantharayan Street. That house was significant to Janaki as it was two houses apart from where she and her husband had lived before Ramanujan left for England. She lived there for fifty years. Her learning in Bombay stood her in good stead in Madras. She began to earn by making clothes and teaching tailoring to girls. Her pension was inadequate and whatever Ramanujan had kept for her was grabbed by unscrupulous relatives.

G.H. Hardy asked S.Chandrasekhar, the astrophysicist, to try to find out a good photograph of Ramanujan next time he was in India. The photograph was necessary for the book Ramanujan: Twelve lectures on subject suggested by his life and work, written by Hardy. In 1937 S.Chandrasekhar met Janaki in Triplicane where he got Ramanujan's passport. With the help of a studio he obtained a good photograph of Ramanujan and sent it to G.H. Hardy. Hardy wrote when he first saw the photograph, 'He looks rather ill, but he looks all over the genius he was'.

Around 1948, Janaki Devi became very affectionate towards a small boy, Narayanan. Narayanan was in hospital with his mother as he was suffering from typhus. Janaki Devi visited him in the hospital. Later, unfortunate Narayanan's parents died. So he went to live with Janaki Devi as her son. She adopted him. She supported herself and her son with her profession of a seamstress and on a meager pension. The University enhanced her pension from twenty rupees per month to one hundred twenty five rupees after World War II. After finishing college education Narayanan began to search for a job. He joined the State Bank of India. Janaki arranged her son's marriage and he had three children. Narayanan was a good companion of his mother and looked after her as long as she was alive.

In 1987, Ramanujan's reputation again came in the lime light as it was his birth centenary year. Our first Prime Minister Jawaharlal Nehru and Nobel laureate physicist Chandrasekhar Venkat Raman were born in the years 14.11.1889 and 07.11.1888 respectively. Ramanujan was compared to J.L.Nehru and C.V. Raman, both of whose centennials were being celebrated at about the sametime. I was astonished at reading about Ramanujan in the Science Reporter of December 1987. I had heard the name of Ramanuja from my grandfather in my childhood. Acharya Ramanuja was born at Sriperumbudur, a place between Kanchipuram and Chennai in the year 1017 A.D. He was a great Vaishnavite saint and a spiritual reformer. But I was ignorant about the genius Ramanujan till 1987.

I have a habit of reading biographies of great persons from my childhood. The habit was formed as my mother used to bring biographies from her school library where she was a teacher. Naturally I was attracted by the biography and the talent of Ramanujan. I began to collect and read articles and books about Ramanujan since 1987. It surprised me how a foreign writer Robert Kanigel can write such a good book about an Indian. I have read the book 'The man who knew infinity' not less than ten times. I have followed the book and inserted many indispensable quotations in my book from the book. A film title : The man who knew infinity, has been made based on the book. Though I had mathematics in B. Sc., I had no idea about the past glory of Indian mathematics. Perhaps this biography of Ramanujan gradually led me to Aryabhata and other ancient Indian mathematicians who taught humankind how to count.

Bruce Berndt, Richard Askey and George Andrews, the three American mathematicians toured various places of India to deliver lectures on the occasion of the birth centenary of Ramanujan. They contributed most to the restoration of Ramanujan's fame. Films were made about his life. Ramanujan Mathematical Society started in 1986. Prime Minister Rajiv Gandhi presented the first copy of "The Lost Note-book" to Janaki Devi when the Narosa Publishing House issued it in Madras.

With the advent of centennial, Janaki Devi began to get attention and respect. Her house became a pilgrimage. She began to get homage from mathematicians who visited Madras. Janaki Devi appeared in a British television special to speak about Ramanujan, 'Letters from an Indian clerk'. A foundation presented her twenty thousand rupees and a monthly pension of one thousand rupees. She asked to create a Srinivasa Ramanujan Trust for awards and scholarships to meritorious mathematics students. Council of Trinity College agreed to grant £2000 a year.

Paul Granlund, the famous American sculptor, made ten bronze busts of Ramanujan based on the passport photograph. It was possible for Richard Askey who pioneered the project and S.Chandrasekhar helped him. Mathematicians around the world contributed to that project. In 1985, in a function the University of Madras presented one of the busts to Janaki Devi; thus her desire for a bust of her husband was fulfilled. The bust stands on a pedestal in Janaki's house in Madras.

Janaki Devi led a life of struggle, however, she got due homage in the last quarter of her life. She breathed her last on April 13, 1994 at the age of 94 years.

Kupuswami Srinivasa Iyengar, Ramanujan's father had gone blind when Ramanujan left for England. There was a high percentage of blindness in India due to lack of adequate treatment facility. In those days cataract was also a cause of blindness in old age. Death of Ramanujan, his dearest son, was unbearable to him. He got sick. He was nursed by Komalatammal and Narayana Iyer's wife. He breathed his last in November, 1920.

Komalatammal broke down mentally and physically after her son's unacceptable death. Her confidence in religious belief and rituals loosened. She again began to face financial hardship. Unlike now-a-days, the then government employees had very poor pay structure. It was very difficult for government employees to maintain their family. In 1927 she wrote a letter to Hardy requesting to intercede for her sons' higher position that they might get an enhanced salary.

Anantharaman, a friend of Ramanujan from childhood heard the news of his friend's death while recovering from a leg operation. Komalatammal even in the late 1930s would visit him and his family in Triplicane to console herself by seeing son's friend as though she were seeing her son Chinnaswami (pet name of Ramanujan)

Dr. P.S. Chandrasekhara, the tuberculosis specialist who had treated Ramanujan in his last months, had the opinion that Ramanujan could have been saved. He wrote in his diary the following day of Ramanujan's death: If he had been allowed to follow my instructions, this double tragedy need not have taken place. The neglect of Ramanujan during his early phase-perhaps partly due to the ignorance of his contemporaries, as well as his relatives' (mother's and wife's) contributory (I almost feel like using the stronger word "criminal") negligence have contributed to this double tragedy - a tragedy which is too deep for tears.

After passing intermediate examination (the new name for the F.A. examination) Lakshmi Narasimhan served for the post office in Triplicane. He died while still young.

Tirunarayanan, the youngest brother of Ramanujan, was born in 1905 when Ramanujan was 17. He continued his study even after the death of his elder brother. He got his B.A degree from the Presidency College. He joined the postal department and became an assistant postmaster.

G.H.Hardy remained alive for about twenty seven years after Ramanujan's death. He always felt the absence of Ramanujan during those surviving years. He was a bachelor and mathematics was his meditation and cognition. After getting to know Ramanujan, he got absorbed in mathematics completely. His wholehearted efforts knew no bounds for the full development of Ramanujan's talent. He devoted most of his time in editing Ramanujan's theorems to make them suitable for publishing in journals. His another effort was to reveal Ramanujan's rare talent to the exclusive circle of mathematicians. Ramanujan's death was one of the cruellest blows to his life. His grief was expressed by the following sentences. "For my part, it is difficult for me to say what I owe to Ramanujan-his originality has been a constant source of suggestion to me ever since I knew him, and his death is one of the worst blows I have ever had."

He was in search of another Ramanujan in various colleges and universities throughout the rest of his life but he had to remain unsuccessful to achieve the desired talent like Ramanujan. Ramanujan was a prodigy, so it was very difficult to find another Ramanujan.

After Ramanujan had sailed for India, Hardy could not concentrate on mathematics in Cambridge. He had got an offer from Oxford University where he joined the New College. He had been in New College just a few months when he received the news of Ramanujan's death from Madras:By the direction of the [University] Syndicate, I write to communicate to you, with feelings of deep regret, the sad news of the death of Mr. S.Ramanujan, F.R.S. which took place on the morning of the 26th April.

After the end of the World War I Germans were boycotted by the people of the other European countries. German mathematicians also were not immune to this type of relegation. G.H Hardy protested against that type of discrimination against German Mathematicians. He had boycotted the mathematical congress held in Strasbourg in 1920 where German, Austrian and Hungarian mathematicians were not invited. In 1921, while Hardy was visiting Germany he wrote to friend Mittag-Leffler: "For my part, I have in no respect modified my former views, and am in no circumstances prepared to take part in, subscribe to, or assist in any manner directly or indirectly, any congress from which, for good reasons or for bad, mathematicians of particular countries are excluded."

After Ramanujan, Littlewood had become the companion of Hardy in his mathematical research. Littlewood was in Cambridge and Hardy was one hundred miles away in Oxford; the only medium for their research were letters. Here Hardy achieved more fame; nevertheless he could not settle here mentally. Hardy joined Princeton University of America for the academic year 1928-1929. He was invited from various universities of America to deliver lectures. He used to deliver lectures and looked for another talent like Ramanujan. Though Hardy was a contented mathematician in America, he made up his mind to return to Oxford as he thought it was more important to work at Oxford.

In1931 he again joined Cambridge University, as Sadleirian Professor, following the death of E.W.Hobson. Now, he had become a famous mathematician. He had been receiving awards for his contribution in mathematics; nevertheless he was not satisfied as Ramanujan was not alive. Articles about Ramanujan were being published in various journals written by Hardy. He wrote about the originality of Ramanujan in one article: He would probably have been a greater mathematician if he had been caught and tamed a little in his youth; he would have discovered more that was new, and that, no doubt, of greater importance. On the other hand he would have been less of a Ramanujan and more of a European professor, and the loss might have been greater than the gain.

Hardy wrote to Subrahmanyan Chandrasekhar from Cambridge on February 19, 1936: I am going to give some lectures (here and at Harvard) on Ramanujan during the summer. Those lectures became the basis of the book- Ramanujan: Twelve lectures on subjects suggested by his life and work. According to a reviewer it was a labor of love.

The three hundredth anniversary of Harvard University foundation was being celebrated with fervor and gaiety. More than two thousand and five hundred celebrities were invited for the Tercentenary Conference of Arts and Sciences. Many Nobel Prize winners were among them.

On the second evening, five lakh enthusiastic people gathered on both the banks of the Charles River to witness two hours of fireworks. Hardy stood before the audience at about nine in the evening on the first day of the conference. The audience before him was not ordinary people but an august group of Nobel scholars. He could deliver a lecture on his invention in mathematics but he started about his best discovery- the talent of Ramanujan: I have set myself a task in these lectures which is genuinely difficult and which, if I were determined to begin by making every excuse for failure I might represent as almost impossible. I have to form myself, as I have never really formed before and to try to help you to form, some sort of reasoned estimate of the most romantic figure in the recent history of mathematics; a man whose career seems full of paradoxes and contradictions, who defies almost all the canons by which we are accustomed to judge one another, and about whom all of us will probably agree in one judgment only, that he was in some sense a very great mathematician. Hardy continued the lecture about Ramanujan.

The creativity of Hardy diminished after a heart attack in 1939 while dusting his bookcase. Then he was sixty two. He could no longer play his favorite games and his mathematical output declined from six to one or two per year.

Hardy was against any war. The beginning of World War II and the waning of his mathematical ability depressed him. But his depression used to disappear when he delivered lectures to his students. He remained a god to his advanced students till his retirement. Hardy retired from the Sadleirian Chair in 1942 at the age of 65. In 1941 while Hardy was watching a rugby match between Cambridge and Oxford a photographer for the British magazine Picture Post snapped his picture. That was a chilly winter day. He was wrapped up in heavy woolen clothes, making him look old. The photograph was disliked by his sister when she saw it.

By 1946, Hardy was virtually an invalid. He became too weak to walk a few yards. His sister Gertrude came to nurse him in Cambridge. Like her brother she also lived in academic settings and never married. She called her beloved brother "Harold". She was not allowed to stay in her brother's room at night as the rules of Trinity were very strict.

The great mathematician Godfrey Harold Hardy died on December 1, 1947 the day he was to be presented the Copley Medal, the highest honor of the Royal Society. With his death the magnificent era of mathematics ended. Hardy and Ramanujan will remain side by side forever in the history of mathematics.

Another talented Indian mathematician was Harish –Chandra (1923-1983), considered to be the greatest Indian mathematician after Ramanujan, was elected F.R.S. in 1973, M.S.Narasimhan and C. Seshadri in 1988, S.R.S.Varadhan in 1998 and M.S.Raghunathan in 2000.

Thirty nine(39) Indian mathematicians and scientists or of Indian origin from different fields were awarded prestigious F.R.S. upto 2000. A total of twenty six (26) F.R.S. were awrded to Indian or Indian origin scientists from 2001-2019. Dr. Gogandeep Kang is the first Indian woman F.R.S. elected in 2019.

1.11 Ramanujan's Mathematics.

Ramanujan's mathematics is incomprehensible to ordinary people and school students. Some of his mathematics are included in the syllabus for the students of college and university level. But his magic squares are interesting and amazing to children. Ramanujan began the first chapter of his notebook with magic squares. Magic squares date back at least to the tenth and eleventh centuries in India. A magic square is merely an array of integers in a square which, when added up by rows or columns yield the same total. The following arrangement he recorded at page two of his notebook 1:

In this magic square the sum of these numbers in a row as well as in a column is 20.

10	2	8
4	5	11
6	13	1

Row
10+2+8=20
4+5+11=20
6+13+1=20

Column.
10+4+6=20
2+5+13=20
8+11+1=20

This magic square is from page one of his notebook 1. The sum of three numbers in a row as well as in a column is 15. Here sum of integers in each diagonal is also 15. The middle term 5 is equal the average of numbers in the row or column or diagonals.

6	1	8
7	5	3
2	9	4

The magic square created by Ramanujan is very interesting. Here magic number is 139. The sum of four numbers in a row, column, diagonal, is 139. Except that sum of four numbers is 139 in various arrangements. The sum of four numbers situated in corners is 139, four numbers situated in the middle part of the square is also 139. If we observe

22	12	18	87
88	17	9	25
10	24	89	16
19	86	23	11

Life of Ramanujan

carefully an interesting fact will come out from the magic square. The numbers in the first row are 22,12,18,87, which is the date of birth of Ramanujan 22.12.1887.

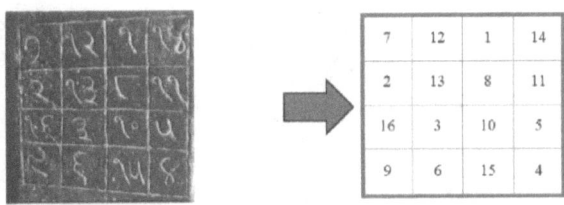

A similar famous and early magic square is inscribed at Parshvanatha Jain Temple at Khajuraho, Madhya Pradesh. This magic square is a proof of Indian advancement in mathematics one thousand years ago. This inscription dating from 954 AD by its builder Pahila, mentioning donation of gardens and requesting future generations to safeguard the temple. This magic square is referred to as the Chautisa (thirty four) Yantra, since each sub-square sums to 34. In this inscribed magic square we can see that zero is very smaller than other digits.

Ramanujan displayed great talent in mathematics even at a young age and had re-discovered many theorems of renowned mathematicians like Euler's Identity: $e^{ix} = \cos x + i \sin x$.

Ramanujan's work covers vast areas including prime numbers, hyper geometric series, modular functions, elliptic functions, mock theta functions, geometry of ellipses etc. the following are three samples :

a) $\sqrt{(1+2\sqrt{(1+3\sqrt{(1+4\sqrt{(1+5(\sqrt{1+...})))))}}}} = 3$;

b) The number 1729 is the minimal natural number having double representation as sum of two cubes : $1729 = 10^3 + 9^3 = 12^3 + 1^3$;

c) $1/\pi = (\sqrt{8}/9801) \sum_0^\infty [\,(4n)!(1103 + 26390n)]$
$/[(n!)^4 \times 3964^n]$:

This series of Ramanujan, which converges at an extraordinarily rapid rate, was applied by others for machine computation of the value of pi to several million digits.

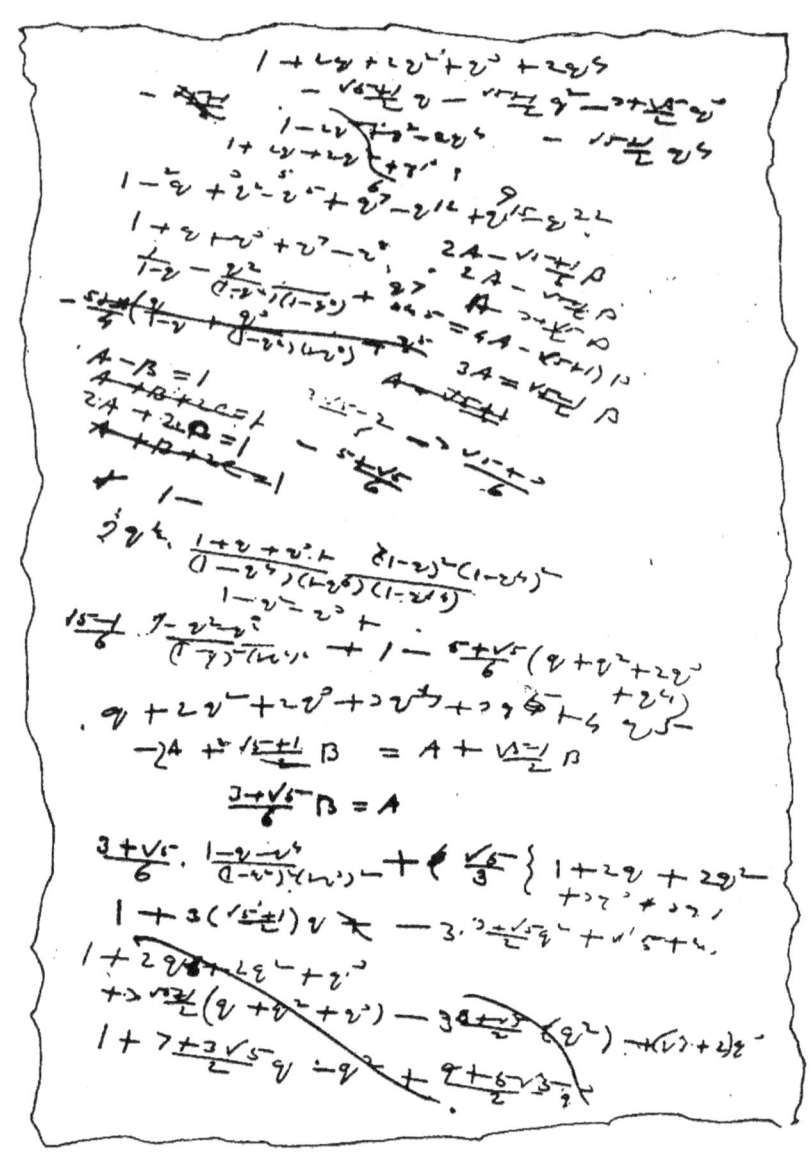

A page of formulas from Ramanujan's 'Lost notebook' discovered by George E. Andrews at Cambridge University.

SQUARING THE CIRCLE

(Journal of the Indian Mathematical Society, v, 1913, 132)

Let PQR be a circle with centre O, of which a diameter is PR. Bisect PO at H and let T be the point of trisection of OR nearer R. Draw TQ perpendicular to PR and place the chord $RS = TQ$.

Join PS, and draw OM and TN parallel to RS. Place a chord $PK = PM$, and draw the tangent $PL = MN$. Join RL, RK and KL. Cut off $RC = \overline{RH}$. Draw CD parallel to KL, meeting RL at D.

Then the square on RD will be equal to the circle PQR approximately.

For $$RS^2 = \tfrac{5}{36}d^2,$$
where d is the diameter of the circle.

Therefore $$PS^2 = \tfrac{31}{36}d^2.$$

But PL and PK are equal to MN and PM respectively.

Therefore $$PK^2 = \tfrac{31}{144}d^2, \text{ and } PL^2 = \tfrac{31}{324}d^2.$$

Hence $$RK^2 = PR^2 - PK^2 = \tfrac{113}{144}d^2,$$
and $$RL^2 = PR^2 + PL^2 = \tfrac{355}{324}d^2.$$

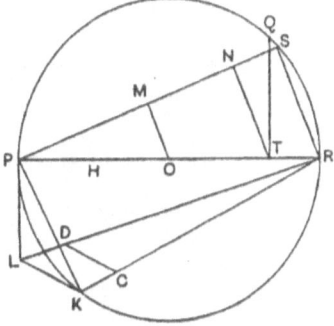

But $$\frac{RK}{RL} = \frac{RC}{RD} = \frac{3}{2}\sqrt{\frac{113}{355}},$$
and $$RC = \tfrac{3}{4}d.$$

Therefore $$RD = \frac{d}{2}\sqrt{\frac{355}{113}} = r\sqrt{\pi}, \text{ very nearly.}$$

Note.—If the area of the circle be 140,000 square miles, then RD is greater than the true length by about an inch.

1.12 Ramanujan's Notebooks

At present, Ramanujan's original three notebooks have been kept in the library of Madras University. Papers and letters related to Ramanujan have been kept in the National Archives in New Delhi and Tamilnadu Archives. Ramanujan's many letters, papers and commentaries by Hardy, Watson and Wilson have been kept in the Wren Library of Trinity College.

We Indians celebrated the year 2012 as the National Mathematics Year of India for the first time. It was the 125th birth Anniversary of Srinivasa Ramanujan. We observe National Mathematics Day every year on 22nd December as it is the birthday of Ramanujan.

Numerals
Of
Different Countries

We, the people of India, have a notion that discovery and invention activities belong only to Europeans and Americans. We have acquired this concept perhaps due to about two hundred years of dependence. The decline of our moral standards and forgetting of our heritage are due to the deep-rooted dependence.

In the middle of the 1920s, two different types of statistics were proposed to describe the behavior of elementary particles known at that time. The first one was proposed in 1924 by Satyendranath Bose and Albert Einstein and the second one was proposed in 1927 by Enrico Fermi and Paul Dirac. It is remarkable that to this day, all the known particles obey the rules dictated by either of these statistics and it is truly strange that there are only two types of statistics and no other to describe a conglomeration of particles. Therefore, all the known particles fell neatly into two groups. Those that obeyed the first were called Bosons and the ones that obeyed the latter were called Fermions. The CERN laboratory in 2012 ultimately detected the elusive Higgs Boson particle which is popularly known as God particle. Peter Higgs and Francois Englert were awarded Nobel Prize in physics in 2013 for this discovery but Satyendranath Bose was not awarded Nobel Prize though his contribution in this field was about ninety years earlier.

Jagadish Chandra Bose (1858-1937) was the greatest interdisciplinary scientist in India. He made remarkable discoveries both in physics and botany working in a very poor research set up. Most of the time he had to make instruments for his research. In 1895, he gave the first demonstration of wireless communication in the presence of the Lt. Governor of Bengal in Calcutta Town Hall. Bose transmitted electromagnetic waves from the lecture hall through intervening walls covering a total distance of 75 feet tripping a relay which threw a heavy iron ball, fired off a pistol and blew a small mine. The millimeter wave length frequency that he used in his experiment in 1895 is the foundation of 5G. He generated electrical radiation at 5mm wave or 60GHz at a time when there was no instruments to measure such frequencies. He was about 120 years ahead of his time. He is now gaining recognition as the father of radio science and semi conductors' technology. Guglielmo Marconi demonstrated trans-Atlantic radio communication in 1901. Marconi got the Nobel Prize in 1909 jointly with Karl Ferdinand Braun for contribution to the development of wireless telegraphy.

The first trans-Atlantic wireless signal in Marconi's experiment was received by him using the iron-mercury-iron coherer with a telephone invented by J.C.Bose in 1898. Bose presented his new invention in a paper to the Royal Society in 1899.

A Nobel Prize has no power to suppress or alter truth. J.C. Bose was the real inventor of wireless and the truth holds him in high esteem.

Fertilization of the ovum by the spermatozoon is the first step in the birth of a baby. For various reasons when fertilization is not possible inside the body, it is performed outside the body and the fertilized ovum is placed in the uterus. The baby born by this process is called a Test Tube baby.

On 25th July 1978, the world's first human test tube baby, Louise Joy Brown, was born in England. The doctors were Robert Edward and Patrick Steptoe. In the methodology followed by them, an ovum was collected using a laparoscope.

But Subhas Mukherjee, a doctor of Kolkata, increased the number of ova using hormones. Without using a laparoscope, he collected the ovum by performing a small operation.

Durga, the first test tube baby in India and the second test tube baby of the world was born on 3 October 1978.

Dr. Subhas's claim was rejected by an incompetent and jealous committee. Instead of recognition he had to face social ostracization, bureaucratic negligence and ridicule. He could not bear the ridicule and committed suicide on 19th June 1981.

Dr. T.C. Anand Kumar developed a test tube baby named "Harsha" (born 16th August 1986). Dr. Kumar was then director of the institute for research in reproduction, Mumbai (a unit of the Indian Council of Medical Research). Despite getting the honor of developing the first human test tube baby in India, he took the pain of going through all the research documents of Mukherjee and was convinced that Mukherjee was the architect of the first human test tube baby in India. On her 25th birthday, Durga revealed her identity for the first time in a ceremony organized in the memory of Dr. Mukherjee. His work got recognition in India and abroad after his death. His method is accepted and used by doctors in all countries.

Like modern times many mathematicians, astronomers, physicians were born in ancient India who were pioneers in their field in the ancient world. Not only India but also Egypt, China, Greece, Babylon and Arabian countries were very advanced in the field of astronomy and mathematics in ancient times.

Ancient people felt the necessity of reckoning to maintain their lives in a proper way. They had their unique instrument on their two hands, the ten fingers. When the full moon would appear in the sky that they could enjoy the moonlight at night and the tide would come to make rivers full to the brims. When the rainy season would come that they could sow seeds for cultivation. They had to reckon everythings in their daily life with the progress of civilization. Astronomy came first in human life. They could predict a season by counting stars or by recognition of a constellation of stars.

At first numbers were written by marks or symbols; gradually due to evolution those became 1,2,3,4,5,6,7,8,9,0 as are written today throughout the world. These numbers were invented by Indian mathematicians and accepted by Arabian merchants and spread by them to European countries.

2.1 Ancient Roman Numeration

We are acquainted with Roman numbers from the first day at our school in childhood. Inscription of the name of a class on the door or on threshold still today is as class I, class II, Class III and so on in Roman Numbers.

Many watches and clocks have inscription in Roman numbers for twelve hours. The basic Roman numbers are as follows

I =1, V=5, X=10, L=50, C=l00, D=500, M=l000

A numeral placed after one of greater value adds to its value:

VI=5+1=6

A numeral placed before one of greater value subtracts from its value:

IV=5-1=4.

Repeating a numeral double its value:

XX=10+10=20

A dash over a number multiplies its value by a thousand:

$\overline{X}$ =10 X 1000 = 10000

Numerals Of Different Countries

Some Roman numbers are given below in the following table.

1	I	19	XIX
2	II	20	XX
3	III	30	XXX
4	IV	40	XL
5	V	50	L
6	VI	90	XC
7	VII	100	C
8	VIII	200	CC
9	IX	400	CD
10	X	500	D
11	XI	900	CM
12	XII	1000	M
13	XIII	5000	$\bar{V}$
14	XIV	10000	$\bar{X}$
15	XV	50000	$\bar{L}$
16	XVI	100000	$\bar{C}$
17	XVII	500000	$\bar{D}$
18	XVIII	1000000	$\bar{M}$

I was a student of Serampore College in West Bengal. It was established by William Carey, Joshua Marshman and William Ward in 1818. It is one of the oldest colleges in India. If I write 1818 in Roman numbers it would be big number 1818 =MDCCCXVIII. Roman numbers were the dominant number system in Italy up to the 13th century and in other countries of Western Europe till the 16th century. Performing arithmetical operations with multi digit Roman numerals was an arduous task. So Indo-Arabic numerals were gradually accepted by Europeans when those numerals reached there by Arabian merchants.

2.2 Ancient Greek Numeration

In ancient Greece the numbers 1,2,3,4 were denoted by vertical strokes I, II, III, and IIII similar to the preBrahmi numerals of ancient India. The numbers one to ten are given in the following table:

I	II	III	IIII	Γ	ΓI	ΓII	ΓIII	ΓIIII	Δ
1	2	3	4	5	6	7	8	9	10
1 - 10 in Greek acrophonic numbers									

The numbers 100, 1000, 10000, were denoted by H, X, M respectively.

From the third century B.C. the denotation of numbers were simplified by using letters of the Greek alphabet. The numbers from 1 to 9 were denoted by the first nine letters of the alphabet. They had 27 letters at that time. The Modern Greek alphabet has 24 letters.

A bar was placed over numerals in order to distinguish them from letters.

Numerals Of Different Countries

	Units	Tens	Hundreds
1	α alpha	ι iota	ρ rho
2	β beta	κ kappa	σ sigma
3	γ gamma	λ lambda	τ tau
4	δ delta	μ mu	υ upsilon
5	ε epsilon	ν nu	φ phi
6	ϝ digamma	ξ xi	χ chi
7	ζ zeta	ο omicron	ψ psi
8	η eta	π pi	ω omega
9	θ theta	ϟ koppa	ϡ sampi

In ancient times alphabetic numeration was used by people of many countries but it is difficult to say who used them first.

Greek mathematics owes its impetus to the influence of its neighboring countries Egypt and Babylon. The ports of the Nile were opened to Greek trade for the first time during the 26th Dynasty of Egypt (C.685-525 BCE). Renowned mathematicians of Greece Pythagoras and Thales visited Egypt to acquire knowledge of mathematics of that country even crossing the Mediterranean Sea by sail-ship, such were their thirst for knowledge. Greeks came in contact with the culture and ideas of Mesopotamia through its neighboring kingdom Lydia.

Alexander the Great conquered many countries up to India (326 BC). His visit to India made a bridge between India and Greece. Indian mathematics reached there but was not accepted at that time as they had their own system of counting.

At that time Chandra Gupta Maurya was the emperor of Northern India. Selucius, a General of Alexander, inherited Alexander's eastern empire from Asia Minor to the Indus after Alexander's departure. He dared to cross his borders and attack Chandra Gupta Maurya. Selucius was defeated and he had to surrender the greater part of his eastern possessions of Gandahar in Afghanistan. The Mauryan Empire then extended from Afghanistan to Bengal and from Kashmir to the Vindhyas.

But a Friendship grew between India and Greece. Chandra Gupta Maurya married the daughter of Selucius. Megasthenis, an ambassador from Greece visited the court of Chandra Gupta Maurya and his description about India is a valuable document of Indian history. Emperor Ashoke, the grandson of Chandra Gupta Maurya sent many Indians to Greece to preach Buddhism.

Greeks accepted Babylonian numeral system with 60 as its base. They also learned to divide circles into 360 degrees. The use of 60 as a base of a mathematical system is unique. 60 is a number that has many divisors (1,2,3,4,5,6,10,12,15,20,30,60) which makes it easier to deal with calculations involving fractions.

The word trigonometry derived from the Greek words trigon and matron and it means measuring the sides of a triangle. We have read Pythagoras theorem in our schools. It states that in a right angled triangle the square erected on the hypotenuse is equal to the sum of the squares erected on the other two sides. It is to be noted that this theorem is stated in terms of geometric objects instead of numbers. Influenced initially by the Egyptians, Greek mathematics made a significant contribution to geometry, astronomy and other subjects related to mathematics.

Plato (C428-348 BC), the great philosopher of Greece had keen interest in mathematics. He founded his Academy in Athens in 387 BC. In his opinion mathematics was a way of understanding more about reality, and geometry was the key to unlocking the secrets of the universe. The sign above his Academy entrance read: "Let no one ignorant of geometry enter here".

Euclid (C.325-C.265 BCE) the Greek mathematician who lived in Alexandria had a good knowledge with all the Greek mathematical work that had been done by his predecessors. He arranged all that knowledge in a single volume. The name of the book is The Elements.

Besides the above mentioned mathematicians many mathematicians lived in ancient Greece, they were Archytas, Archimedes, Chrysippus, Democritus, Diony and many others.

2.3 Slavic Numeration

The Slavic people of the south and east of Europe used the alphabetic system of notation for writing numbers. The letter used as a numeral was surmounted by a special symbol placed over numerals. Slavic numeration persisted in Russia till the end of the 17th century. Then Indo-Arabic numeration was accepted in Russia. The Slavic numerals:

2.4 Ancient Armenian and Georgian numeration

Armenians and Georgians used the alphabetic principle of numeration. Alphabets of these people had more letters than that of Greeks. So they used special symbols for the numbers 1000, 2000, 3000, 4000, 5000, 6000, 7000, 8000 and 9000. The numerical values of the letters followed the order of the letters in the alphabets of these people. Slowly Indo-Arabic system of numeration replaced their system. In Armenia the alphabetic numeration is still used in designations of chapters, stanzas etc.

2.5 Babylonian Numeration

The ancient Babylonians developed a positional system or place value system as early as the 19th century BC but the Babylonian systems were based on 60. Later Indian sages developed a base 10 positional systems or decimal system which is used still now throughout the world.

Numbers less than 60 were denoted by two symbols: ⟨Y⟩ for unity and ⟨⟨⟩ for ten. These signs were repeated as many times as required, for example:

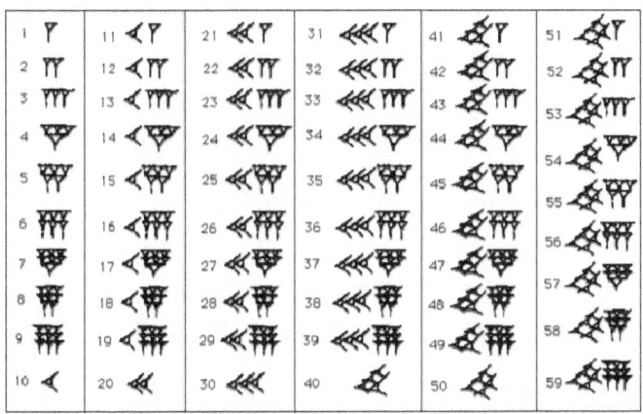

Babylonian symbols of whole numbers did not spread beyond the Assyrian-Babylonian Empire, but sexagesimal fractions spread to many countries of central Asia, Northern Africa and Western Europe. We still use the ancient Babylonian sexagesimal fractions in the divisions of the degree of angle and arc, and the hour of time into 60 minutes, and of the minute into 60 seconds.

2.6 Ancient Mayan Numeration

The Western Hemisphere's greatest civilization was developed between AD 250 and 900 by the Maya people who lived in Central America, occupying the area which today is southern Mexico, Guatemala and northern Belize. The Maya existed as far back as 2000 BC. Over the centuries they developed their land by draining marshy land and building irrigation system. From 300 BC to AD 300, they built many cities in Guatemala, Belize and South Yucatan. Their cities were well planned. They built their buildings, temples, pyramids, palaces, market places with their artistic style. The Maya people laid out their sacred shrines and temples covering many hectares of land, with large open spaces, platforms and meeting places.

The prime time of Maya civilization was between AD250 and 900. The Mayan hieroglyphic (picture writing) writing are found on stone monuments and written on books made of bark papers. Pages of their books were arranged like a bellows of an accordion.

By AD 665 they used a place-value number system to base 20 with a symbol for zero. Mayan astronomy used an accurately determined solar year and precise tables of the positions of the Venus and the Moon. Accurate calendar was important for elaborate rituals and ceremonies of the Maya religion.

The zeros used by the Mayans in their books are identifiable as shells and are printed red. Their method of expressing numbers with zero was different than that of Indian.

Numerals Of Different Countries 121

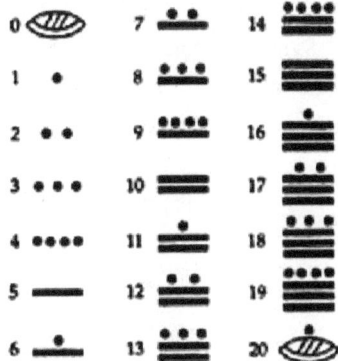

It is a pity that this great achievement and innovation in mathematics by Maya civilization did not influence people of any other country of the world. After AD 900 the civilization declined rapidly for unknown reasons.

2.7 Ancient Chinese Numeration

China was very advanced in mathematics in ancient times. Ancient Chinese numeration was older than Brahmi numeration of India. Chinese numerical inscriptions were found on oracle bones of 14th century BC by archeological excavation. The Shang (Shang dynasty of 14th century B.C) numerals for the numbers were:

During Han Dynasty (2nd century B.C.- 4th Century A.D) the numeral system had evolved into computational algorithms carried out with a counting board and set of rods. The numerals and their computing rod configurations are,

I	II	III	IIII	IIIII	T	TT	TTT	TTTT
1	2	3	4	5	6	7	8	9

—	=	≡	≣	≣	⊥	⊥	⊥	⊥
10	20	30	40	50	60	70	80	90

Perhaps ancient mathematicians of China were the first inventor as well as user of magic squares for the construction of dikes, canals etc.

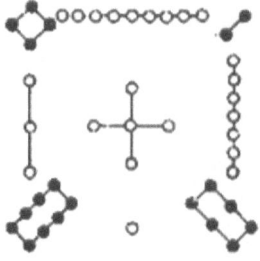

4	9	2
3	5	7
8	1	6

China is one of the oldest (8000 B.C) civilizations comparable only to Egypt, Babylonia and India. Between CA 600-CA 1200 Buddhist and Indian influences were at their best. In the wake of Buddhism, Indian culture and science were followed. China and India now shared their knowledge but the Indian knowledge was not used for mathematics, rather for astronomy.

They used blank space where we would write 0, for expressing a number. 65023: T≣   =III

Liu Hui was one of the most brilliant mathematicians in

human history. He calculated value of π as 3.14 10 24. Zu Chongzhi (429-501AD) and Zu Geng (450-520), father and son were great astronomers. Zu Chongzhi calculated a near accurate value for π and worked for a new calendar. Between 628 and 656 mathematics was established at the imperial academy. Mathematician Zhen Luan lived at the end of the 6th century.

The first circular symbol for zero in China is found in the book of Qin Jiushao (1247). His answer to the subtraction of 1470000 - 64464 = 1405536 is written as |≡ O ≡ ||||| ≡ T

However zero was already in use during the Tang dynasty (626-907AD) for one of the Indian scholars, Qutan Xita. He wrote a dot instead of an empty space, giving it the same function as the zero symbol. This may indicate that the zero in China originated from India. Golden era of Chinese mathematics was in the 13th and the beginning of the 14th century. Four great mathematicians Qin Jiushao, Yang Hui, Li Zhi (also named Li Ye) and Zhu Shijie lived during that period and enriched Chinese mathematics.

Cheng Dawei completed his systematic Treatise on Arithmetic in 1542 AD, in which extensive use of abacus was made to solve problems. Abacus (the Chinese suan-pad) was introduced in about 1200 C.E. Abacus was developed in Mesopotamia in 1000 BC from a flat sand-covered stone counting board on which pebbles were moved.

In ancient time there was no problem of passport visa and dreadful diseases. So students, traders and travellers had access to move from one country to another freely.

Three scholars from China visited India in ancient time. Their written description about India is a valuable source of ancient Indian history. All students of India know about them as their names and activities are included in school syllabus.

Faxian (337-422 CE): Faxian was a Chinese Buddhist monk. He is also known by the names Fa-Hien and Fa-hsien. He travelled by foot all the way from China to India crossing icy desert and rugged mountain passes of North-West India. Faxian visited Pataliputra during the reign of Chandragupta II. He visited many sacred places of Buddhist sites in India (undivided India including Pakistan and Bangladesh), Nepal and Sri Lanka between 399 and 412 with the intention of collecting Buddhist scriptures. After returning to China by sea route he translated some works which he had collected in India. The description of his journey and record of Buddhist Kingdoms are valuable documents of ancient India.

Xuanzang (602-664) : Xuanzang was also a famous Chinese Buddhist monk. He was ordained as a novice monk at the age of thirteen and a Bhiksu (full monk) at the age of twenty. He knew about Faxian's visit to India and, like him, contemplated to visit India to acquire complete Buddhist texts. In 629, Xuanzang dreamt a dream that convinced him to journey to India.

Xuanzang passed through modern Kyrgyzstan, Uzbekistan and Afghanistan and entered into India through Khyber Pass. During his travel he visited many Buddhist sites and relics including Buddhas of Bamiyan carved out of rock. In India, he visited Taxila (now in Pakistan),where he found 5000 more Buddhist monks in 100 monasteries. Between 632 and early 633, he studied with various monks. Later he visited Kanyakubja (Kannauj), the magnificent capital of emperor Harsha. There he met 10000 monks in 100 monasteries. Xuanzang stayed in the city to study early Buddhist scriptures. After visiting Bodh Gaya he reached Nalanda University, the greatest ancient Indian university, where he spent two years in the company of several thousand scholar-monks to study logic, grammar, Sanskrit and Buddhist scriptures. He met venerable scholar Silabhadra, the monastery's superior, who was to make

Numerals Of Different Countries

known to Xuanzang the ultimate secrets of the idealist systems. He returned to China in 645 AD with over 600 texts. In 646, Xuanzang completed his book Great Tang Records on the Western Regions. His description of ancient India is a valuable historical document.

Yijing (635-713): Yijing was a Chinese Buddhist Monk and a traveller. He came to India through sea-route and duration of travel was 25 years. The written records of his travel give us information of the ancient kingdoms of India and Nalanda Buddhist University. The duration of his stay at Nalanda University was eleven years. He translated a large number of Buddhist texts from Sanskrit into Chinese. After completing all translation works he returned to China by sea-route in the year 695. He brought back about 400 Buddhist translated texts.

The above mentioned Scholars remained in India to acquire Buddhist scriptures for long years. It is obvious that they learned at least elementary Indian mathematics and carried the knowledge to China though we have no historical evidence.

Traders from China used to come to India to sell their products. They brought abacus with them for calculation. Chinese traders were influenced by Indian simple digits with their individual values. Gradually they accepted Indian simple method of writing numbers. Ancient China presented the humankind an art which is the pioneer of civilization.

The earliest form of hand printing technology was developed in China, Japan and Korea. Books in China were printed by rubbing paper against the inked surface of wood blocks since AD 594. The imperial state in China was the major producer of printed material for a long time. Buddhist missionaries from China introduced hand printing technology into Japan around

AD 768-770. Marco Polo, the great explorer, returned to Italy after many years of exploration in China, in 1295. Marco Polo brought with him the knowledge of wood-block printing technology to Italy. Now, Italians began producing books with the help of that technology and soon the technology spread to other parts of Europe. Johann Gutenberg of Germany developed the first printing press. He made metal types for the letters of the alphabet. Gutenberg perfected his system by 1448. The first book he printed was the Bible. He printed about 180 copies in three years. It was a fast production in respect of that time.

Paper is essential for printing and it was also invented in China. Ts'ai Lun (50-121), a courtier in the Chinese imperial court invented paper in 105. He mixed tree fibers and wheat stalks with the bark of a mulberry tree, then pounding them together and pouring the mixture into a woven cloth to create a lightweight writing surface. His blended, fibrous sheets were an improvement over bamboo and wood which were awkward. This invention was kept secret by successive Chinese dynasties. Meanwhile paper making technology appeared in Japan and Korea from the beginning of 7th century.

Arabian Soldiers captured Chinese paper merchants during the Battle of Talas in 751, as a result of that capture knowledge of paper making spread across the Arab World. From there the knowledge of paper making appeared in Europe early in the 12th century and parchment lost its throne.

2.8 Ancient Egyptian Numeration

We have read in our geography book that Egypt is the gift of the river the Nile and it is a country where colossal pyramids are situated.

We know about Thuten Khamen from childhood as he and the other Pharaohs are described in our history book. Egyptian settlement along the fertile Nile valley was one of the oldest in the world. They settled there as early as about 6000 BC. Egyptians were attracted by night sky and had got the natural calendar there. They began to record the patterns of lunar phases as well as patterns of star constellation for seasons. Idea of season was necessary for them, both for agricultural and religious purposes.

A base 10 numeration system developed in Egypt as early as 2700 BC. To denote a number they used a stroke for units, a heel bone symbol for tens, a coil of rope for hundreds and a lotus plant for thousands; other symbols for higher powers of ten up to a million. It is illustrated below :

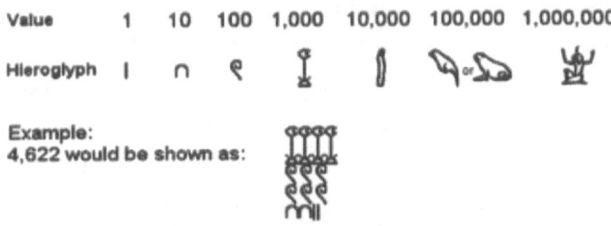

Hieroglyphics for Egyptian numerals

Egyptians had no concept of place value, so larger numbers were clumsy. However they had knowledge of arithmetic and geometry. Egyptians had clear conception of fraction. Following example explains it.

If they needed to divide 3 loaves among 5 people, they would first divide two of the loaves into thirds and the third loaf into fifths then they would divide the left over third from the second loaf into five pieces.

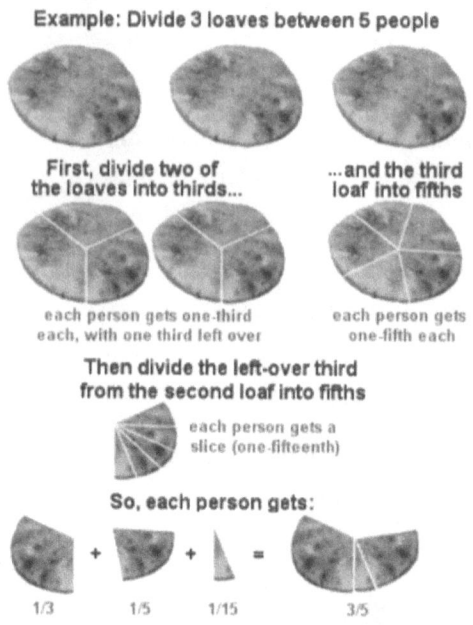

For measurement the Pharaoh's surveyors used body parts. Palm, hand, elbow to finger tips etc. were used for that purpose. Repeated flooding of the Nile washed away markers that separate one person's land from that of others. Geometry originated in Egypt for measurement of land over and over again to demarcate altered configuration of the landscape. It would have been impossible for the architects of Egypt to build colossal pyramids and other surprising monuments without the knowledge of mathematics and mensuration.

Mathematicians of Egypt observed that the area of a circle of diameter 9 units, for example, was very close to the area of a square with sides of 8 units. The area of circles of other diameters could be obtained by multiplying the diameter by 8/9 and then squaring it. It is certain that they knew the formula for the volume of a pyramid: $1/3 \times$ Length $\times$ breath $\times$ height. Architects of Egypt knew that a triangle with sides 3, 4 and 5 units yields a perfect right angle, long before Pythagoras. Egyptian architects used ropes knotted at intervals of 3, 4 and 5 units in order to ensure exact right angles for their construction of pyramids and other monuments with stones. The 3-4-5 right triangle is often called "Egyptian".

2.9 Indian Numeration

Indian Numeration: Our forefathers used to lead a very simple life and high thinking. Our ancestors who were sages evolved a numeral system, which we call the Decimal System. It was a creative endeavor on the basis with only ten symbols-they should be able to represent any number as well as they have their individual value. It is very simple to us now-a-days but only a brilliant brain could have developed the system. It is a pity that we do not know their names or it is not possible to bring their names from the depth of history.

The ten numerals or digits 0, 1,2,3,4,5,6,7,8,9 are the basis for calculation or writing any number even today.

Our forefathers made another big invention by positioning or placing these symbols. The far right position in any numeral became the unit. Moving left, the next position became ten. Thus we get hundreds, thousands and so on by moving left. Positions are written as follows:

1(one in unit position)
10(one in tenth position)
100(one in hundredth position)
1000(one in thousandth position)

Let us take a complex number : the price of a mobile phone as 5087.

If we break it in decimal system then

$5087 = (5 \times 1000) + (0 \times 100) + (8 \times 10) + (7 \times 1)$

I have written several lines to explain positional system or decimal system but our ancient mathematicians explained it more clearly in a Sanskrit shloka (verse). Which is stated below:

"एकदशशतसहस्रायुतलक्षप्रयुतकोटयः क्रमशः ।
जलधिश्चान्त्यं मध्यं परार्धमिति दशगुणोत्तरं संज्ञाः।
संख्यायाः स्थानानां व्यवहारार्थं कृताः पूर्वैः॥१०-११॥"

[Līlā-KVS, p-8]

Chanting of Sanskrit verses is very melodious.

Sanskrit is an old Indo-Aryan language. The most ancient form is in Rig Veda, dating from the late 2nd millennium BC. Panini codified the Sanskrit language in a grammar in 5th century BC. Our ancient mathematicians composed difficult mathematical formulas in simple verses for everyone's comprehension.

The above mentioned Sanskrit verse states that: Positions of the digits from right to left are unit, ten, hundred, thousand, ten thousand, hundred thousand (lakh), million, ten million (crore), hundred million, billion, kharva, nikharva, mahapadma, sanku, jaladhi, antya, madhya, parardha. The value of each digit on the left is ten times that on the right. For this simplicity the Indian numeral system was accepted by the people of all the countries of the world and is used even today.

In my childhood I saw in my village Beliatore that cooking was performed in every house by burning wood in a small hole on the earthen floor of kitchen. Utensils were earthenware. Women from forest area used to bring heaps of small bundles of fire wood from the jungle. They were not paid regularly. They got payment at the end of month. So they used to draw vertical lines by a chalk on the wall of the kitchen according to the number of bundles of wood supplied on that day. Suppose on a particular day four bundles of wood were supplied to a house, the stroke on the wall would be IIII, if five bundles were supplied next day, five vertical lines would appear by the side of the previous lines, IIIII. Illiterate milk-men also used the same process. They used to draw vertical line according to one fourth of a seer. They could not count or understand more than twenty (20).

If a person was asked, "Could you tell me your age?" He would said, "Two twenty and six". That is 46 years. Perhaps their perception of the twenty grew owing to total number of fingers on our two hands and two legs.

In my maternal uncle's house at Baidyabati, I saw in my childhood some people from Odisha (then Orissha) used to supply drinking water from nearby tube wells. Even Bengalis were not allowed to supply drinking water to Bengali families as people from Odisha were regarded as sanctified as they were from Lord Jagannatha's place. They also used one vertical stroke for two buckets of water which they wrote on a door. Washer men from Bihar or UP also used one vertical stroke for a cloth and got payment at the end of month. Though they were illiterate but counting was necessary to lead their daily life or to earn their livelihood.

I saw writing of vertical lines by the people of Bengal, Bihar, Odisha and UP about fifty five years ago. It could not be seen now-a-days due to increase of literacy. Perhaps it was an ancient system of counting in India and an old indigenous numerical notation, where the nine numbers were represented by the corresponding number of vertical lines. Perhaps vertical line system was the origin of the Brahmi numerals.

1	2	3	4	5	6	7	8	9
I	II	III	II II	III II	III III	IIII III	IIII IIII	III III III

Ifrah's guess for pre-Brahmi numbers

Brahmi numbers were much in vogue in India before the birth of Jesus Christ. Those numbers were modified version of vertical lines to be written rapidly in order to save time. But there were no special symbols for 2 and 3, both numbers being constructed from the symbol for 1, which was a horizontal line.

1	2	3	4	5	6	7	8	9
—	=	≡	+	h	4	7	ら	?

Brahmi numerals around 1st century A.D.

As there was not a place value system so there were separate Brahmi symbols for 10, 20, 30, 40, 50, 60, 70, 80, 90, 100, 300, 400 etc.

The Brahmi numerals have been found in inscriptions in caves and on coins excavated in regions near Pune, Mumbai and in some places in Uttar Pradesh. Dating these numerals tells us that they were in use over quite a long period span up to the 4th century A.D.

Brahmi numerals were somewhat different in different parts of India. During the reign of Gupta dynasty in the Magadha state in northern India from the early 4th century AD to the late 6th century AD, the Gupta numerals developed from the Brahmi numerals and were spread over large areas by the Gupta Empire.

1	2	3	4	5	6	7	8	9
—	=	≡	૪	ᖾ	♃	η	ᔕ	ꓛ
Gupta numerals around 4th century A.D.									

Here the system of writing 1, 2 and 3 in Gupta numerals remained same as that of Brahmi numerals which were horizontal lines.

The Gupta numerals evolved into Nagari or Devanagari numerals, the later name literally means the 'Writing of the Gods'. Nagari form evolved from the Gupta numerals beginning around the 11th century AD and continued to develop from the 11th century onward.

1	2	3	4	5	6	7	8	9	0
٩	२	३	८	५	६	७	८	९	०
Nagari numerals around 11th century A.D.									

According to al-Biruni, Nagari numerals were the best and most regular figures in India. Nagari numerals had been transmitted into the Arab World, by his time. Here Zero was presented by a full round circle.

An important discovery

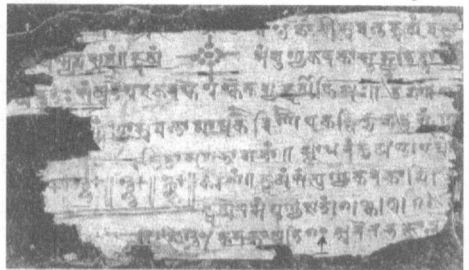

A page from the Bakhshali manuscript with a dot in the last line (arrow marked) used as a placeholder for zero.

The manuscript was first discovered by a local farmer in 1881, and was named after the village it was found in, near Peshawar now in Pakistan. The manuscript consists of 70 leaves of birch bark, filled with mathematics and text in the form of Sanskrit. The oldest pages are from somewhere between 224 AD and 383 AD. It is kept in the Bodleian Library of Oxford University since 1902.

On a charter found at Sankheda in Gujarat, the year is written in decimal notation as '346', the year corresponding to 594-6 AD.

2.10 History of Zero

It is very difficult to answer a question: Who discovered Zero? The conception of zero which we have now-a-days was a gift to us by our forefathers who were saints and sages around 400 BC. Babylonians put two hooks symbols into the place where we would put zero to indicate a difference between two hundred sixteen, and two thousand one hundred six, 216 and 21"6. Mesopotamians used three hooks to denote an empty place in the positional notation around 700 BC. Clay tablets which were excavated from present Iraq also contain single hook for an empty place.

Ptolemy, the 2nd century Egyptian astronomer who saw the Earth as the center of the universe with Sun, Moon and the five known planets moving around it. His theory is known as the Ptolemaic or geocentric system. Around 130AD he used Babylonian sexagesimal system together with the empty place holder 0. He used 0 both between digits and at the end of a number but Ptolemy considered it as a sort of punctuation mark.

Though some historians wanted to play down Indian contribution in number system and concept of zero but it is unreasonable.

In around 500 AD Aryabhata devised a number system which has no zero yet was a positional system. He used Indian alphabet "Kha" for position and it would be used later as the name for zero. In ancient India dots also had been used to denote an empty place in positional notation.

We the Indians have a stone tablet in Gwalior in Madhya Pradesh, which has an inscription dated 933 in the Vikrama Calendar, in our calendar it is 876.

Where a garden was planted, 187 by 270 (according to some other's opinion it can be 210)hastas which would produce enough flowers to allow 50 garlands per day to be dedicated to the Deity of the local temple. Both the numbers 270 and 50 are denoted almost as they appear today although the 0 is smaller and slightly raised.

We have more ancient zero in our ancient text called Bakhshali manuscript as a prominent dot as a place holder for zero. The oldest pages of this manuscript are from between 224AD and 383 AD.

Brahmagupta, the ancient Indian mathematician, first formulated the concept of zero around 628 AD. At the beginning, zero was not considered as a number which was used to denote an empty space. The use of zero as a number came into Indian mathematics around 650 AD. In AD 628 Brahmagupta wrote his book *"Brahmasphutasiddhanta"* and gave rules for zero and negative numbers.

He explained:

Any number - that same number =0

Any number + 0 = that same number

Any number - 0 = that same number

Any number × 0 =0

Any number ÷ 0=0

His last rule division by zero was not correct.

Another Indian Mathematician Mahavira wrote his book "Ganita Sara Samgraha" (Collection of Mathematics Briefings) in 830 AD. It was an update of Brahmagupta's book Brahmasphutasiddhanta. He also gave incorrect rule for division by zero.

Later another mathematician Bhaskara II (1114-1185) stated that square of zero is zero and square root of zero is also zero which are correct. But he stated that any number divided by zero is infinity. Conceptually it is well but it is still incorrect.

Zero is a unique number not only in mathematics but also all branches of sciences would have struggled for more clear definitions had zero not existed in our number system.

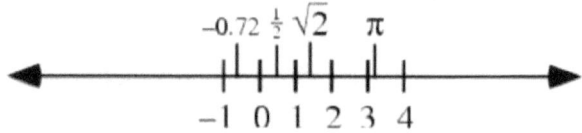

Zero is just like a pivot as it occupies the middle position of a number line. Zero is neither positive nor negative, it is neutral. If we move to the right side from zero, the numbers are positive. If we move to the left side of zero, we get negative numbers. Zero has no power of its own. But if we start writing it on the right side of a number then it begins to show its power and the number increases by ten times for each additional zero.

I have mentioned earlier that according to Bhaskara II if any number is divided by zero, the answer is infinity but it is not correct. If we continue to divide a real positive number by smaller and smaller numbers the result will go on increasing. For example:

$10/0.01 = 1000$
$10/0.0001 = 100000$
$10/10^{-99} = 10^{100}$ and so on.

Whatever may be the smaller number but still the smaller number is not equal to zero. Moreover, infinity is a concept, an abstract thing, not a number as defined in our number system. All rules of mathematics are invalid with infinity. For example, if we add infinity to infinity, the result is not twice the value of infinity. It is still infinity.

If we divide 20 by 4 we get 5. Again if we multiply 5 with 4, we get original value 20 back again. Applying algebra, we can write it as follows:

If $a/b = c$, then $a = b \times c$
Let us take, $a = 20, b = 0, c = \infty$ (infinity)
Then, $20/0 = \infty$ (infinity)
Or $20 = 0 \times \infty$ (infinity)

We know anything multiplied with zero is zero.

So 20=0, this is an absurd result. We cannot get back 20 by multiplying the digits.

Therefore, it is wrong to say that a number divided by zero is infinity. It is absurd to attempt to divide a number by zero. So division by zero is undefined for real numbers.

Another case is zero divided by zero. An expression like zero divided by zero is defined as indeterminate. Let us take 0/0=10, 1000 or any number. The rule of multiplication also holds true here since 10 or 1000 multiplied by zero will give the product as zero. Nevertheless, we cannot determine the exact or precise value for this expression. So 0/0 is said to be indeterminate. Similarly zero to the power zero is also indeterminate.

Mighty Zero: We known one hundred $=100=10^2$

And one thousand $=1000=10^3$ but billion is different in the USA and the UK. In the USA and France, a billion equals a thousand million, where as in England and Europe except France a billion is a million million. The following table will make our confusion clear.

Numerical values across the world

Number	USA/France	UK/Europe (Except France)	India
10^5	One hundred thousand	One hundred thousand	One lakh
10^6	Million	Million	Ten lakh
10^7	Ten million	Ten million	One crore
10^8	Hundred million	Hundred million	Ten crore
10^9	billion	Milliard (Thousand million)	Hundred crore
10^{12}	Trillion	Billion (million million)	
10^{15}	Quadrillion	Thousand billion	
10^{18}	Quintillion	Trillion	
10^{21}	Sextillion	Thousand trillion	
10^{24}	Septillion	Quadrillion	
10^{27}	Octillion	Thousand quadrillion	
10^{30}	Nonillion	Quintillion	
10^{33}	Decillion	Thousand quintillion	

Etymology of zero: In Indian languages zero is called Shunya which means empty or nothing. AI-Khwarizmi, the famous mathematician of Iraq, translated Shunya as "Al-saifor" or Saifar, later it became Zephyr in Latin. Afterwards the word Zephyr became Zero.

Big Bang and big numbers: In Indian philosophy Creator of the Universe and his creation (the universe) was confined in a very small space, which we may call shunya or zero position in respect of this vast universe. Then the Almighty, the creator of this universe decided to create this universe. Thus his total concentrated energy was distributed into all Anu (molecules) and Paramanu (atoms) of the universe. The galaxies, stars, planets, later lives were formed from chemical compounds .Thus almighty became infinity from his almost zero existence, as He is present in every atom of the universe, so He is omnipresent. He is present in living and non-living objects. This philosophy exists in India from time immemorial.

It may be a coincidence but it is an astonishing fact that the Indian philosophy has been corroborated by modern physics.

We know about Big Bang theory. Before Big Bang there was no space and no time, that is, no galaxy, no star and no planet. The Big Bang is the gradual expansion and maturation of the universe. It began 13.7 billion years ago and continues through the present day. To explain the idea clearly, here I insert a few lines from three books written by three renowned scientists.

Professor Stephen W. Hawking: As the big bang itself, the universe is thought to have had zero size, and so to have been infinitely hot. But as the universe expanded, the temperature of the radiation decreased. One second after the big bang, it would have fallen to about ten thousand million degrees. This is about a thousand times the temperature at the center of the sun, but temperature as high as this are reached in H-bomb explosions. At this time the universe would have contained mostly photons, electrons, and neutrinos (extremely light particles that are affected only by the weak force and gravity) and their antiparticles, together with some protons and neutrons. As the universe continued to expand and the temperature to drop, the rate at which electron/antielectron pairs were being produced in collisions would have fallen below the rate at which they were being destroyed by annihilation. So most of the electrons and antielectrons would have annihilated with each other to produce more protons, leaving only a few electrons left over. The neutrinos and antineutrinos, however, would not have annihilated with each other, because these particles interact with themselves and with other particles only very weakly. So they should still be around today. If we could observe them, it would provide a good test of this picture of a very hot early stage of the universe. Unfortunately, their energies nowadays would be too low for us to observe them directly. However, if neutrinos are not massless, but have a small mass of their own, as suggested by an unconfirmed Russian experiment performed in 1981, we might be able to detect them indirectly: they could be a form of "dark matter" like that mentioned earlier, with sufficient gravitational attraction to stop the expansion of the universe and cause it to collapse again.

Dr. Robert L. Piccioni: We know with very good precision that the universe began 13.7 billion years ago. Our universe began as a fantastically small object with a fantastically high temperature. It may have been as small as the Planck length, less than a millionth of a millionth of a millionth of a millionth of the size of the smallest atom. And its initial temperature was probably over 100 million, million, million, million, million degrees. The universe has been expanding and cooling ever since.

Dr. Mani Bhaumik: There is much evidence to be gathered before we have the full picture. Lucky for us, nature has left much of the evidence 'hidden in plain sight' throughout the cosmos. The trail of this evidence leads us grippingly close to the very beginning, when science tells us that all that we see in the starry sky and all that lies beyond it came from a seed tantalizingly far, far smaller than the full stop at the end of this sentence.

Mathematics was part and parcel of Indian religions namely Hinduism, Buddhism and Jainism. American scientist **Carl Sagan** explained Hindu religious faith and modern astronomical idea nicely in his book Cosmos. Here I insert two pages from his book:

Carl Sagan

Every human culture rejoices in the fact that there are cycles in nature. But how, it was thought, could such cycles come about unless the gods willed them? And if there are cycles in the years of humans, might there not be cycles in the aeons of the gods? The Hindu religion is the only one of the world's great faiths dedicated to the idea that the Cosmos itself undergoes an immense, indeed an infinite, number of deaths and rebirths. It is the only religion in which the time scales correspond, no doubt by accident, to those of modern scientific cosmology. Its cycles run from our ordinary day and night to a day and night of Brahma, 8.64 billion years long, longer than the age of the Earth or the Sun and about half the time since the Big Bang. And there are much longer time scales still.

There is the deep and appealing notion that the universe is but the dream of the god who, after a hundred Brahma years, dissolves himself into a dreamless sleep. The universe dissolves with him until, after another Brahma century, he stirs, recomposes himself and begins again to dream the great cosmic dream. Meanwhile, elsewhere, there are an infinite number of other universes, each with its own god dreaming the cosmic dream. These great ideas are tempered by another, perhaps still greater. It is said that men may not be the dreams of the gods, but rather that the gods are the dreams of men.

In India there are many gods, and each god has many manifestations. The Chola bronzes, cast in the eleventh century, include several different incarnations of the god Shiva. The most elegant and sublime of these is a representation of the creation of the universe at the beginning of each cosmic cycle, a motif known as the cosmic dance of Shiva. The god, called in this manifestation Nataraja, the Dance King, has four hands. In the upper right hand is a drum whose sound is the sound of creation. In the upper left hand is a tongue of flame, a reminder that the universe, now newly created, will billions of years from now be utterly destroyed.

These profound and lovely images are, I like to imagine, a kind of premonition of modern astronomical ideas. *Very likely, the universe has been expanding since the Big Bang, but it is by no means clear that it will continue to expand forever. The expansion may gradually slow, stop and reverse itself. If there is less than a certain critical amount of matter in the universe, the gravitation of the receding galaxies will be insufficient to stop the expansion, and the universe will run away forever. But if there is more matter than we can see - hidden away in black holes, say, or in hot but invisible gas between the galaxies- then the universe will hold together gravitationally and partake of a very Indian succession of cycles, expansion followed by contraction, universe upon universe, Cosmos without end. If we live in such an oscillating universe, then the Big Bang is not the creation of the Cosmos but merely the end of the previous cycle, the destruction of the last incarnation of the Cosmos.

*The dates on Mayan inscriptions also range deep into the past and occasionally far into the future. One inscription refers to a time more than a million years ago and another perhaps refers to events of 400 million years ago, although this is in some dispute among Mayan scholars. The events memorialized may be mythical, but the time scales are prodigious. A millennium before Europeans were willing to divest themselves of the Biblical idea that the world was a few thousand years old, the Mayans were thinking of millions, and the Indians of billions.

Persons those who have read Stephen Hawking's A Brief History of Time are acquainted with Carl Sagan. Introduction of this book was written by Carl Sagan.

2.11 Propagation of ancient Indian Mathematics

The Arabs, because of their extraordinary mobility propagated Indian mathematics to different parts of the world. Arabian merchants came to India from times immemorial for trade and commerce. Gradually they accepted Indian counting system as it was easy to express big numbers as well as addition, subtraction, multiplication were quick with Indian digits with their absolute value and unique zero. We know people of all countries of the world had their own system of counting and mathematics for calculation to meet the demand of daily life and to enhance their own progress.

Acceptance of Indian system was not smooth and quick. None wants to abandon his country's heritage and custom, it is just like degradation. Arabians accepted Indian mathematics without any hesitation but it took centuries to be acquired by Europeans.

Priceless mathematical manuscripts created by our ancient mathematicians could not be preserved due to foreign invasion and to some extent of our negligence. However we are lucky enough that those manuscripts were translated by Arabian mathematicians and preserved in their countries.

In 773 an Indian scholar visited the court of the king of Baghdad Kholof-al-Mansoor. He carried with him a copy of an Indian astronomical text, quite possibly Brahmagupta's Brahmasphutasiddhanta. The Caliph ordered to translate the work in Arabic. The Caliph was very interested in mathematics and all branches of science. He sent some persons to Sindh (Now in Pakistan) for studying mathematics, astronomy and medicine. After completing their study they carried copies of important books. Later these books were translated to Arabian languages.

AI-Khwarizmi: One of the best and multifarious talented Arabian mathematician was Abu Ja'far Muhammad ibn Musa al-Khwarizmi. He is popularly known as Al-Khwarizmi (780 AD -850 AD). His name perhaps indicate that he came from Khwarezm now in Uzbekistan. AI Khwarizmi was a Persian mathematician, astronomer and geographer during the Abbasid Caliphate. He was a scholar in the House of Wisdom in Baghdad.

AI-Khwarizmi

AI-Khwarizmi was most impressed by the efficiency of Indian numeral system. He popularized the Indian numeral system through his book, Kitab al-Zam Wa-1-tafriq bi-hisab al-Hind (The Book of Addition and Subtraction According to the Hindu Calculation). This book was written by him in 825 AD.

Here Hindu implies not to Hindu religion or Hinduism, but to the people living beyond Hindu Kush mountain range. Ancient Indian mathematics was developed by mathematicians belonging to Buddhism, Jainism, Hinduism, etc. living at that time in India.

This is the earliest available arithmetic text that deals with Indian numbers. A Latin version of this book was done by Adelard of Bath in Europe in the twelfth century but no extant Arabic manuscript is available. The Latin manuscripts are commonly referred to by the first two words with which they start: Dixit algorizmi (so said al-Khwarizmi), or Algoritmi de numero lndorum

(al-Khwarizmi on the Hindu Art of Reckoning)

Al-Khwarizmi's work on arithmetic played a crucial role for introducing the Indian numerals to the Middle East and Europe.

Algebra: Al-Khwarizmi was a genius and contributed much in all branches of mathematics. His systematic approach for solving linear and quadratic led to Algebra. The word Algebra was derived from the title of his book on the subject, Al-Kitab al-mukhtasar fi hisab al-jabr wa-1-muqabala. (The Compendious Book on Calculation by Completion and Balancing). This book was written by him in 830 AD. The term Algebra is derived from the term al-jabr, the name of one of the basic operations with equations. Al-jabr means restoration referring to adding a number to both sides of the equation to consolidate or cancel terms.

Caliph al-Mamun was very enthusiastic about mathematics, geography and astronomy. The books were written by him with the encouragement of the Caliph al-Mamun. His Algebra book is full of examples and applications to a wide range of problems in trade, surveying and legal inheritance.

The book was translated in Latin as Liber algebrae et almucabala by Robe of Chester in 1145. The book was translated in English by F.Rosen in 1831. An Arabic copy is kept in Oxford and a Latin translation is kept in Cambridge.

Astronomy: Al-Khwarizmi wrote a book on astronomy, Al-Khwarizmi's Zij al-Sindhind, in 820 AD. This book consists about 37 chapters on calendrical and astronomical calculations and 116 tables with calendrical, astronomical and astrological data, as well as a table of sine values. This book has been written based on the Indian astronomical methods known as the Sindhind.

Al-Khwarizmi's work marked the turning point in Islamic astronomy. His work contains tables for the movements of the sun, the moon and the five planets Mercury, Venus, Earth, Mars and Jupiter known to humankind at that time.

The original Arabic version is lost. A version by Maslamah Ibn Ahmad al-Majriti, a Spanish astronomer has survived, in a Latin translation by Adelard of Bath.

The four surviving Latin translations are kept at the following places.

>Bibliotheque Publique (Chartres)
>Bibliotheque Mazarine (Paris)
>Biblioteca Nacianal (Madrid) and
>Bodleian Library (Oxford)

Trigonometry: Trigonometry is part and parcel of astronomical calculations. AI-Khwarizmi's Zij al-Sindhind contains tables for the trigonometric functions of sines and cosines as well as spherical trigonometry.

Geography: AI-Khwarizmi had multifarious intelligence. Though he was a mathematician, geography was a subject of interest for him. His geography book is Kitab Urat al-Ar (Book on the appearance of the Earth or The image of the Earth). He finished this major work in 833. This geography book is a revised and completed version of Ptolemy's geography. It consists a list of 2402 coordinates of cities and other geographical information. A Latin translation is kept at the Biblioteca Nacional de Espana in Madrid. Only one copy of Kitab Urat al-Ar is kept at Strasbourg University Library.

AI-Biruni: Full name of al-Biruni was Abu Rayhan Muhammad ibn Ahmad AI-Biruni. He was born on 4/5 September in 973 at Khwarezm now in

AI-Biruni

Uzbekistan. Later his birth place was named Biruni to honour him.

AI-Biruni is regarded as one of the greatest scholars of the medieval Islamic era. He had keen interest in all subjects physics, anthropology, comparative sociology, astronomy, astrology, chemistry, history, geography, mathematics, medicine, psychology, philosophy, theology. He learned many languages and was well conversant with Khwarezmin, Persian, Arabic, and Sanskrit and also knew Greek, Hebrew, and Syrian.

AI-Biruni was made court astrologer by Mahmud of Ghazni in 1017. He accompanied Mahmud on his invasions into India and lived there for a few years. He became the most important interpreter of Indian science to the Islamic World. He was given the title al-Ustadh (The Master) for his remarkable description of early eleventh century India.

AI-Biruni became acquainted with all things related to India. He learned Sanskrit to go through Indian books. He wrote one of the major works Kitab ta'rikh al-Hind, finishing it around 1030. An extensive description on Indian astronomy, mathematics, religion, history, geography, science was written in this book. The lunar crater AI-Biruni is named in his honor. He died on 13th December, 1048.

Omar Khayyam: Omar Khayyam was born on 18 May 1048, in Nishapur in North Eastern Iran. He died on 4th December 1131 and was buried in Nishapur. He was a talented mathematician, astronomer, philosopher and poet. At his young age he went to Samarkand and obtained his education there. After completing his education he moved to Bukhara where he established himself as one of the major mathematicians and astronomers of that period.

Omar Khayyam is the author of the Treatise on Demonstration of Problems of Algebra. This important treatise includes a geometric method for solving cubic equations by intersecting a hyperbola with a circle.

The Panchangam (the Hindu calendar) was devised by Aryabhata and his followers have been in use in India for the practical and religious purposes.

By day Omar Khayyam would teach algebra and geometry and at night as he was an astronomer would study astronomy to complete important aspects of the Jalani Calendar.

The Jalani Calendar was introduced by Omar Khayyam with other mathematicians and astronomers in 1073 CE. Modified versions (modified in 1925) of this Jalani Calendar are the national calendars in use in Iran and Afghanistan. The dates of the Jalani calendar are based on actual solar transit, as in Aryabhata and earlier Siddhanta calendars.

Leonardo Fibonacci: Leonardo Fibonacci (C.1170-C.1250) was an Italian mathematician. He is considered the most talented western mathematician of the Middle Ages.

Fibonacci realized that arithmetic with Hindu-Arabic numerals is simpler and more efficient that with Roman numerals. He travelled several Arabian countries for studying mathematics under the renowned Arabian mathematicians of the time. After completing his study he returned in 1200 AD and began composing a book. His book Liber Abaci (Book of Abacus or Book of Calculation) was completed in 1202.

Leonardo Fibonacci took the initiative to popularize Hindu Arabic numerals in Europe by his book Liber Abaci. He recommended

Leonardo Fibonacci

Indian numeration with the digits 0 to 9 and place value system. He showed in his book the practical importance of the Indian numeral system by applying it to commercial book keeping, conversion of weights and measures, calculation of interest and so on. Liber Abaci became widely accepted throughout Europe. The book could make a profound impact on European thought.

I have mentioned three Arabian and one Italian mathematician but many other mathematicians have contribution for propagating Indian mathematics. If I write that 'India is the Mother of Mathematics' it would not be exaggeration. Genius mathematician like Ramanujan was born in recent times and talented mathematicians like Aryabhata, Bhaskara were born in ancient times in India. As for mathematics at least five thousand basic and advanced modern mathematical concepts have their root in India and most of these have Vedic antecedents. Moreover ancient counting system of India is the counting system of modern world. Someone may think that I am an Indian so I am trying to prove or establish more than the truth or fact.

But many renowned mathematicians and scientists like Albert Einstein paid homage to the genius of our ancient mathematicians.

The importance of the positional number system of India is described by the French mathematician Pierre Simon Laplace (1749-1827) who wrote:

It is India that gave us the ingenious method of expressing all numbers by the means of ten symbols, each symbol receiving a value of position, as well as an absolute value; a profound and important idea which appears so simple to us now that we ignore its true merit, but its very simplicity, the great ease which it has lent to all computations, puts our arithmetic in the first rank of useful inventions; and we shall appreciate the grandeur of this achievement when we remember that it escaped the genius of Archimedes and Apollonius, two of the greatest men produced by antiquity.

Famous mathematician Tobias Dantzig (1884-1956) who was born in present day Latvia, expressed gratitude to our ancient unknown mathematicians for their discovery of positional number system which we the people of the world use still today. He wrote: This long period of nearly five thousand years saw the rise and fall of many civilizations, each leaving behind it a heritage of literature, art, philosophy, and religion. But what was the net achievement in the field of reckoning, the earliest art practised by man? An inflexible numeration so crude as to make progress well-nigh impossible, and a calculating device so limited in scope that even elementary calculations called for the services of an expert...... Man used these devices for thousands of years without making a single worthwhile improvement in the instrument, without contributing a single important idea to the system.......Even when compared with the slow growth of ideas during dark ages, the history of reckoning presents a peculiar picture of desolate stagnation. When viewed in this light, the achievements of the unknown Hindu, who sometime in the first centuries of our era discovered the principle of position, assumes the importance of a world event.

Albert Einstein's statement regarding our ancient mathematicians is very exceptional. He remarked that, 'We owe a lot to the Indians who taught us how to count, without which no worthwhile scientific discovery could have been made.'

Indian numerals and mathematics, by radically simplifying computations, gave a significant boost to business, international trade and the sciences, and there by making a major contribution to world civilization.

2.12 Evolution of Indo-Arabic numerals

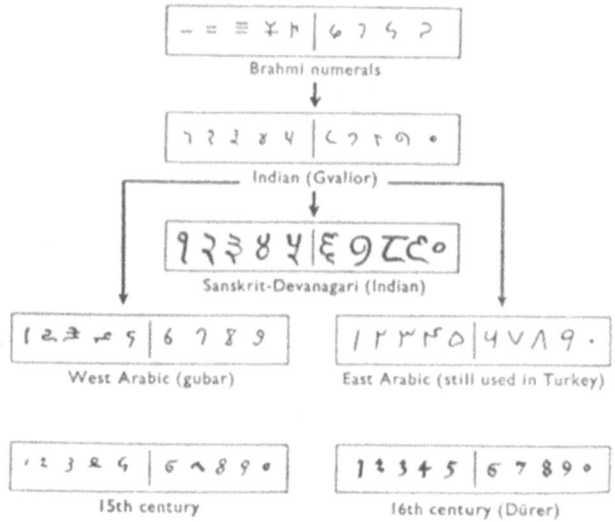

Ancient Indian Mathematicians

Ancient Indian Mathematicians

As are the crests on the heads of peacocks, as the jewels on the hoods of cobras, so is mathematics at the top of all sciences.

— The Yajurveda, 600 BCE

The oldest known mathematics texts in existence are the Sulbasutras of Baudhayana, Apastamba and Katyayana of Vedic age. The Sulbasutras had been estimated to have been composed around 800 BC, but the mathematical knowledge existing in India is much more ancient(3000BC). The Sulbasutras had been used for the constructions of the elegant Vedic fire-altars with precision. The altars had rich symbolic significance and were constructed depicting a falcon in flight with curved wings, a chariot-wheel with spokes or a tortoise with extended head and legs. Ancient Indians thought that it would be a sin if altars are not made with precision and religious rituals are not performed in accurate time. For this reason geometry, trigonometry and astronomy were included in Indian religions namely Hinduism, Jainism and Buddhism from times immemorial. Our ancient people had an idea of a geometric theorem which is known as 'Pythagoras theorem' much before Pythagoras (C 580-500 BC), the Greek philosopher and mathematician was born. They used to apply this theorem in various geometric constructions.

3.1. Baudhayana (800-about 740BC):

Baudhayana was one of the oldest mathematicians of ancient India (India was divided in 1947 so here India implies including Pakistan and Bangladesh). He was connected with Yajurveda School and was the author of the Baudhayana Sutras, which cover dharma, daily rituals, Vedic sacrifices etc. He authored 'Sulbasutra'-an addendum to the 'Vedas' giving rules for the construction of altars.

Altars were made with great precision requiring mathematical calculation. Ancient Indian mathematical erudition was composed entirely in verses. Difficult sutras could be easily memorized and the knowledge got orally transmitted to successive generations. This system was a boon as paucity and perishability of writing materials could not stop propagation of knowledge.

Whatever Baudhayana discovered in ancient times, that was 2800 years ago and mind-boggling to us the modern people.

(a) He gave us a method of constructing a circle equal in area of a square.

(b) His value of $\sqrt{2}$ is not much different from modern values.

$$\sqrt{2} \approx 1 + \frac{1}{3} + \frac{1}{3 \times 4} - \frac{1}{3 \times 4 \times 34} = \frac{577}{408} \approx 1.414216$$

(c) Baudhayana was the first mathematician in the world who had calculated the value of Pi (π). He calculated value of

$$\pi = \frac{900}{289} = 3.114$$

Srinivasa Ramanujan in 1914 gave an approximate value for $\pi = 3.1415926525826461252..........$

(d) Baudhayana Theorem (Pythagoras Theorem): We Indians should call Pythagoras Theorem as Baudhayana Theorem as it was discovered by him at least two hundred years before Pythagoras (C 580-500 BC) was born.

Baudhayana described this theorem in his book Baudhayana Sulbasutra. His book is also one of the oldest books on mathematics. The shloka (verse) that describes Baudhayana theorem is stated below:

दीर्घचतुरश्रस्याक्ष्णया रज्जुः पार्श्वमानी तिर्यग् मानी च
यत् पृथग् भूते कुरूतस्तदुभयं करोति ॥

The translation of the shloka (verse) is,

A rope stretched along the length of the diagonal produces an area which the vertical and horizontal sides make together.

Baudhayana used a rope as an example to make the theorem comprehensive to everyone. Ancient Indian mathematicians described their mathematical results in simple shlokas (verses). A few sets of squares are given in the Sulbasutra:

$$3^2+4^2=5^2, 5^2+12^2=13^2, 7^2+24^2=25^2, 8^2+15^2=17^2$$

$$12^2+16^2=20^2, 12^2+35^2=37^2$$

Though this theorem is called Pythagoras theorem he was not the inventor of it. Greek mathematician Sir Tomas Heet said "No really trustworthy evidence exists that it was actually discovered by him".

Dr. Bibhuti Bhusan Dutta said "Instances of application of the theorem occur in the 'Baudhayana' and 'Satapatha Brahman' (X.2.3.7-14). There are reasons to believe it to be as old as the 'Taittiriya' and other samhitas".

3.2. Manava (750BC-about 690BC):

Manava was interested in mathematics using it for religious purposes. He was an author of geometric text of Sulbasutras. Manava was a Vedic priest and he wrote the Sulbasutra to provide rules for religious rites. He was also a skilled craftsman and was able to construct accurate altars needed for sacrifices.

Manava's geometric Sulbasutra describes constructions of circles from rectangles, and squares from circles. He calculated value of $\pi = 25/8 = 3.125$

3.3 Apastamba (600 BC - about 540 BC):

Apastamba was from a family of Brahmins dedicated to the study of the Yajurveda. His Sulbasutra contains principles of geometry needed for Vedic rituals. His work consisted of six chapters.

3.4 Panini (520 - about 460 BC):

Panini was born in a place named Shalatula on the bank of river Indus, now in Pakistan. We know him as the founder of the Sanskrit Grammar but he was also a great mathematician. As he was a mathematician his Grammar work Ashtadhyayi was written in a perfect scientific way. I have already mentioned that Ramanujan's formulas are being used in Computer Science and there are great similarities between the computer programming languages and the Panini's grammar.

3.5 Katyayana:

Katyayana perhaps lived in the North-West region of India in third century BC. He was a priest, Sanskrit grammarian and mathematician. His work Sulbasutra contains 102 sutras (formulas). He was an expert altar builder. Katyayana discussed the geometrical proportions, the different measures, relative positions and special relations of the various vedis (altars) in connection with their constructions. Perhaps he was the first person who gave a description of the measuring tape. His other work was Sanskrit grammar, which was an elaboration on Panini grammar.

3.6 Aryabhata I (476-550 AD):

Some people may think that birth of genius mathematician Ramanujan in India was a rare occurrence, it is not correct to have such a notion. Many genius mathematicians were born in India in ancient times, Aryabhata I is one of them. He was a mathematician and astronomer of great talent.

Scholars have different opinion regarding the birth place of Aryabhata I. Some historians think that he was from Pataliputra or Kusumpura whose present name is Patna in Bihar. Some scholars give the opinion that he was born somewhere in Kerala, South India. Some have the opinion that he was born in present day Maharashtra State. Though scholars have different opinion regarding the birth place of Aryabhata I, they are unanimous that his sphere of activity was at Pataliputra.

According to many experts he was the Kulapa (Vice-Chancellor) of the illustrious Nalanda University of ancient India. His work Aryabhatiya is of immense importance. He completed his work in AD 499, when he was 23 years old. Aryabhatiya is a treatise in Sanskrit on mathematics and astronomy.

In this work Aryabhata I did not use the Brahmi numerals. He continued tradition from Vedic times and used letters of the Sanskrit alphabet to denote numbers, expressing quantities in mnemonic verses. He used an ingenious system to express numbers on the decimal place value model.

He gave us the rules for extraction of square and cube roots by the arithmetical method which are rules used all over the world even today.

We cannot imagine the extent of scientific inventions by Aryabhata I if he had a modern day telescope. He was the first Indian astronomer to state that the Earth is spherical and rotates on its axis. He explained that the apparent daily east-west motion of the sun, moon, planets, and stars is due to the rotation of the earth from West to East. In the same way that someone in a boat going forward sees unmoving objects on the banks going backward. He calculated the time of one sidereal rotation of the Earth as 23h 56m 4.1s. The modern value is 23h 56m 4.091s, the difference is very small.

His work Aryabhatiya consists of 108 verses and 13 introductory verses, and is divided into four padas or chapters.

1) **Gitika pada (13 Verses):**

    It is also called Dasagitika or the ten Giti Stanzas. For writing a number Aryabhata used the expression Sthanat Sthanam dasagunam syat (from place to place each is ten times the preceding). It is based on a place –value notation. He expressed revolutions of heavenly bodies based on place-value or decimal system. In a yuga the revolutions of the Sun are 4320000, of the Moon 57753336, of the Earth eastward 1582237500, of the Saturn 146564, of Jupiter 364224, of Mars 2296824, of Mercury and Venus the same as those of the Sun.

2) **Ganitapada or Mathematics (33 Verses)**

    The highest number actually used by Aryabhata himself runs to ten places. But India had at least to the eighteenth places. The numbers used by Aryabhata are as follows. The numbers eka [one], dasa [ten], sata [hundred], sahasra [thousand], ayuta [ten thousand], niyuta [hundred thousand], prayata [million], koti [ten million], arbuda [hundred million], and urnda [thousand million] are from place to place each ten times the preceding.

Ancient Indian Mathematicians

Aryabhata defined square and cube as follows.

The area of a square, and the product of two equal quantities are called Varga. The product of three equal quantities and a solid which has twelve edges are called Ghana (cube).

He gave us methods for calculating square root and cube root of numbers.

a) In the following example the sign ° indicates the varga places, and the sign – indicates the avarga places.

```
                          °⁻ ⁻°⁻°
                          1 5 1 2 9  (root = 1)
Square of the root        1
                          ─
Twice the root            2)05(2 = quotient (or next digit of root)
 (2×1)                      4
                          ───
                           11
Square of the quotient      4
                          ───
Twice the root            24)72(3 = quotient (or next digit of root)
 (2×12)                      72
                          ───
                            09
Square of the quotient       9
                          ───
                             0
             Square root is 1 2 3
```

b) In the following example the sign ° indicates the Ghana places and the sign – indicates the aghana places.

```
                                    °⁻ ⁻°⁻ ⁻°
                                    1 8 6 0 8 6 7  (root = 1)
Cube of root                        1
                                    ─
Three times square of root          3)08(2 = quotient (or next digit of
 (3×1²)                               6        root)
                                    ───
                                     26
Square of quotient multiplied        12
 by three times the pūrva           ───
 (2²×3×1)                           140
Cube of quotient                      8
                                    ─────
Three times square of root          432)1328(3 = quotient (or next digit
 (3×12²)                                1296       of root)
                                         ───
                                         326
Square of quotient multiplied            324
 by three times the pūrva                ───
 (3²×3×12)                                27
Cube of quotient                          27
                                         ───
                                           0
             Cube root is 1 2 3
```

161

Aryabhata bestowed on us the following formulas for calculating the (a) area of a triangle, (b) area of any plane figure, (c) area of a circle (d) area of a trapezium.

a) Area of a triangle: the area of a triangle is the product of the perpendicular and half the base.

$$Area = \frac{1}{2} \times a \times h$$

b) Area of any plain figure: The area of any plain figure is found by determining two sides and then multiplying them together.

$$Area = a \times b$$

c) Area of a circle: Half of the circumference multiplied by half the diameter is the area of a circle.

$$Area = \frac{2\pi r}{2} \times \frac{2r}{2} = \pi r^2$$

d) a and b are the parallel sides of a trapezium and c is the perpendicular between them.

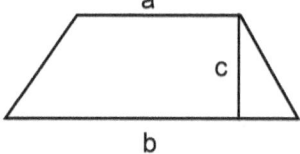

The area of a trapezium is obtained by multiplying half the sum of the two sides (parallel sides) by the perpendicular

$$Area = \frac{(a+b)\,c}{2}$$

3) Kalakriya or Reckoning of time: A year consists of twelve months. A month consists of thirty days. A day consists of sixty nadis. A nadi consists of sixty vinadikas.

At that time division of a day was not like ours hour, minute, second. But name of the days in a week were according to the heavenly bodies which are used till now. The order of the planets is Saturn, Jupiter, Mars, the Sun (a star), Venus, Mercury and the Moon (a satellite). Therefore they become rulers of the days of the week as follows:

Saturday – Saturn	Wednesday – Mercury
Sunday – Sun	Thursday – Jupiter
Monday – Moon	Friday – Venus.
Tuesday – Mars	

4) Gola or the sphere: Translation of some of his verses from this chapter are mentioned below.

Verse No 1. From the beginning of Mesa to the end of Kanya is the northern half of the ecliptic. The other half from the beginning of Taulya to the end of Mina is the southern half of the ecliptic. Both deviate equally from the equator.

Verse No. 7. Just as a ball formed by a Kadamba flower is surrounded on all sides by blossoms just so the Earth is surrounded on all sides by all creatures terrestrial and aquatic.

Verse No. 9

अनुलोमगतिर्नौस्थः पश्यत्यचलं विलोमगं यद्वत्।
अचलानि भानि तद्वत् समपश्चिमगानि लङ्कायां ॥ ९ ॥

As a man in a boat going forward sees a stationary object moving backward just so at Lanka a man sees the stationary asterisms moving backward (westward) in a straight line.

Verse No. 20. The east and west line and the north and south line and the perpendicular from the zenith to nadir intersect in the place where there observer is.

Aryabhata I gave a method for calculating the value of π in a verse.

"चतुरधिकं शतमष्टगुणं द्वाषष्टिस्तथा सहस्राणाम् ।
अयुतद्वयविष्कम्भस्यासन्नो वृत्तपरिणाह॥१०॥"

100 plus 4, multiplied by 8, and added to 62000; this is the nearly approximate measure of the circumference of a circle whose diameter is 20000.

Thus, $\pi = \dfrac{\text{circumference}}{\text{diameter}} = \dfrac{(100+4) \times 8 + 62000}{\text{diameter}} = \dfrac{62832}{20000} = 3.1416$

This value does not occur in any earlier work on mathematics and forms an important contribution of Aryabhata I. It is noteworthy that he has called the above value approximate.

Aryabhata I is the author of several treatises on mathematics and astronomy, some of which are lost. The mathematical part of the Aryabhatiya covers arithmetic, algebra, plane trigonometry and spherical trigonometry. It also contains continued fractions, quadratic equations, sums of power series, and a table of sine.

The Arya-siddhanta, a lost work on astronomical computations, is known through the writing of Aryabhata's contemporary Varahamihira and later mathematicians and commentators including Brahmagupta and Bhaskara I.

A third text, which may have survived in the Arabic translation, is Al nft or Al-nanf.

Aryabhata gives the area of a triangle by the verse.

'त्रिभुजस्य फलशरीरं समदलकोटीभुजार्धसंवर्ग: ॥६॥'

The English translation of this verse is: The product of the perpendicular (dropped from the vertex on the base) and half the base gives the measure of the area of a triangle.

Aryabhata narrated sine by the name of 'ardha-jya' which literally implies half-chord. It became 'jaib in Arabic translation. Later in the 12th century this term became sinus in Latin translation. Later it became sine in English.

Aryabhata presented us three excellent mind boggling formulas for the summation of series of squares and cubes.

$$1 + 2 + 3 + \ldots + n = \frac{n(n+1)}{2}$$

$$1^2 + 2^2 + \ldots + n^2 = \frac{n(n+1)(2n+1)}{6}$$

$$1^3 + 2^3 + \ldots + n^3 = (1 + 2 + \ldots + n)^2$$

We are to remember that he formulated the above equations about one thousand five hundred years ago when digits were not in vogue and counting was in a crude form in all countries of the world.

Ancient people had a belief that eclipse was caused by pseudo-planetary demons Rahu and Katu. Aryabhata states that the moon and planets have no light of their own, they sine by reflected sunlight. He states the lunar eclipse occurs when the moon enters into the Earth's shadow and a solar eclipse occurs when a new moon passes directly between Earth and the sun. His computation regarding eclipses and the positions and periods of the planets are very accurate and differ a few seconds from modern calculation by advanced sophisticated instruments. He established an observatory in the sun-temple situated in Terigona, Bihar. Aryabhata had a belief that the planets' orbits are elliptical. He stated that the planets really all move at the same speed. The nearer ones seem to move more rapidly than the more distant ones because their orbits are small.

Aryabhata was a mathematician unparalleled even in modern times and his discoveries in astronomy were about one thousand years before the birth of Nicolaus Copernicus (1473-1543), Galileo Galilei (1564-1642) and Johannes Kepler (1571-1630)

Various steps have been taken in independent India to honour this great mathematician. First unmanned Earth Satellite built by our country was launched from the erstwhile Soviet Union by a Russian-made rocket on April 19, 1975. It was named after him.

The astronomical research institution situated at Nainital is named after him, Aryabhata Research Institute of Observational Science. The Inter School Aryabhata Maths Competition bears his name. The bacteria invented by scientists of ISRO in 2009 are given a name to honour him, Bacillus Aryabhata.

3.7 Varahamihira (505-587 AD):

Varahamihira was one of the nine jewels (navaratnas) of the court of legendary king Vikramaditya of Malwa, Madhya Pradesh. He lived in Ujjain.

Varahamihira was an astronomer, mathematician, and astrologer. His main work is Pancha-Siddhantika dated 575 AD. This work is a compilation of five earlier astronomical treatises namely the Surya Siddhanta, Romaka Siddhanta, Paulisa Siddhanta, Vasishtha Siddhanta, and Pitamaha Siddhantas. He was the first person to mention that the shifting of the equinox is 50.32 seconds.

Another work of Varahamihira is the Bhirat-Samhita. It covers many subjects including astrology, planetary movements, eclipses, rainfall, clouds, architecture, and manufacture of perfume, matrimony, growth of crops, domestic relations, gems, pearls, and rituals.

3.8 Brahmagupta (598-660AD):

Brahmagupta lived at Bhinamala during the reign of King Vyaghramukha. This village is in Rajasthan situated in between Mount Abu and river Luni. His father's name was Jisnu Gupta and grand-father's name was Bishnu Gupta.

When he was only thirty years of age i.e. in 628, he wrote Brahmasphuta Siddhanta. This work consists of 24 chapters and contains 1008 slokas (verses). Brahmagupta recorded Astronomy, Arithmetic, Algebra and Geometry in this work. In 655, he wrote another book Khandakhadyaka. This work, related with Astronomical texts, has eight chapters.

An Indian scholar visited the court of al- Mansur in Baghdad in 773 AD. He carried with him a copy of Brahmasphuta Siddhanta.The Caliph ordered this work translated into Arabic. Muhammad al-Fazari, the eminent Arabian mathematician performed the translation of the Indian text.

Brahmagupta was the head of Ujjain observatory. He was the first mathematician to introduce zero as a digit in his first work.

Algebra: Some of the equations solved by Brahmagupta are mentioned below.

(a) $ax+c=bx+d$

(b) $x+y=a, x-y=b$

(c) $ax^2+bx=c$

(d) $nx^2+1=y^2$

Arithmetic: Brahmagupta mentioned four methods of multiplication in his work Brahmasphuta Siddhanta. They are goumutrika, khanda, bheda, ista.

(a) Goumutrika: The Sanskrit word goumutrika means similar to the course of cow's urine hence zigzag.

Example: multiplication of 1223 by 235, the numbers are written thus,

```
2        1 2 2 3
3        1 2 2 3
5        1 2 2 3
```

After all the horizontal lines have been multiplied by the corresponding numbers on the left in the vertical line, the numbers stand thus,

```
   2446
   3669
   6115
 _____
 287405
```

After being added together as with present method, the result is obtained.

(b) Khanda method means parts multiplication method. There are two methods under this system. In first method the multiplier is broken up into two or more parts and is then multiplied severally by these and the result is added.

For example:
$$11 \times 156 \quad (11 = 5+6)$$
$$= (5 \times 156) + (6 \times 156)$$
$$= 780 + 936$$
$$= 1716$$

In second method multiplier is broken up into two or more parts. The multiplicand is then multiplied by one of these, the resulting product by the second thus till all the parts are exhausted. The ultimate product is the result.

For example:

56 × 236		(56 = 2 × 4 × 7)
56 × 236	=	(2 × 4 × 7) × 236
	=	(2 × 236) × 4 × 7
	=	472 × 4 × 7
	=	(472 × 4) × 7
	=	1888 × 7
	=	13216

These methods of multiplication are found among the Arabs and the Italians.

(c) Ista method is of two kinds, which is shown by means of examples.

(a) 94 × 13 = (94 + 6) × 13 − 6 × 13
= 1300 − 78
= 1222

(b) 94 × 13 = (90 + 4) × 13
= 90 × 13 + 4 × 13
= 1170 + 52
= 1222

Geometry: Brahmagupta gave an exact formula for the area of a cyclic quadrilateral. The area of a cyclic quadrilateral with sides a, b, c, d, and semi perimeter s is:

$$\text{Area} = \sqrt{(s-a)(s-b)(s-c)(s-d)}$$

(Brahmagupta did not mention the word cyclic)

Astronomy: Brahmagupta gave us methods for calculating the position of heavenly bodies over time (ephemerides), their rising and setting and the calculation of solar and lunar eclipses. He explained that the degree of the illuminated part of the Moon depends on the relative positions of the Sun and the Moon, and this can be computed from the size of the angle between the two bodies. Brahmagupta observed that the Earth and other heavenly bodies are spherical. He had an idea about gravity. He wrote, a body falls towards the earth as it is the nature of the earth to attract bodies just as it is the nature of the water to flow.

3.9 Bhaskara I (600-680 AD):

We had two mathematicians with the same name. Bhaskara II was a 12th century mathematician and I would discuss about him later.

Bhaskara was born at Bori in Parbhani district of Maharashtra. He got astronomical education from his father who was also an astronomer. He dared to deviate from tradition and he was the first to use the Brahmi numerals in the Indian decimal system with a circle for the zero. He used prose in his work. His commentary on Aryabhata's work, Aryabhatiya bhasya written in 629AD, is the oldest known prose work in Sanskrit on mathematics and astronomy. He became famous by writing commentary on Aryabhata's work. Bhaskara's other two works are Mahabhaskariya and Laghubhaskariya.

In his work Mahabhaskariya, in Chapter 7, he gave a remarkable approximation formula for $\sin \theta$,

that is $\sin \theta = \dfrac{16\theta(\pi-\theta)}{5\pi^2 - 4\theta(\pi-\theta)}$, $(0 \leq \theta \leq \dfrac{\pi}{2})$

Ancient Indian Mathematicians

About one thousand and four hundred years ago Bhaskara I gave a formula by which we get very close value of Sin θ in comparison with present value. Percentage of error is only +.00160. A table is stated below:

	sin θ By Bhaskara's formula	sinθ Real value
0	0.00000	0.00000
10	0.17525	0.17365
20	0.34317	0.34202
30	0.50000	0.50000
40	0.64183	0.64279
50	0.76471	0.76604
60	0.86486	0.86603
70	0.93903	0.93969
80	0.98461	0.98481
90	1.0000	1.00000

He stated that if P is a prime number, then 1+ (P-1) is divisible by P. It is known as Wilson's theorem.

Bhaskara stated a problem: tell me, O mathematician what is that square which multiplied by 8 becomes, together with unity, square. In modern notation the equation is $8x^2+1=y^2$.

It has solution pairs x=1, y=3 or x=6, y=17 etc.

In modern times it is called Pell equation. Important parts of his works were translated into Arabic.

3.10 Sridharacharya (870-930AD):

He was born in Bhurishresti village in South Radha (at present Hubli, Karnataka). His father's name was Baladevacharya and mother's name was Acchoka. According to the opinion of some historians, Sridharacharya was born in Bengal while other historians believe that he was born in South India.

Sridharacharya wrote a book named Vyaktaganita similar to BhaskaracharyaII's Lilavati. He wrote another book Ganitasara by name. This work contains the theory of Arithmetic and Mensuration. His other famous book is Trisatika; this work contains 300 slokas (verses) regarding Arithmetic, Algebra and Geometry.

3.11 Bateswar:

In 880 AD, Bateswar was born at Anandpur in North Gujarat. His father was Mahadatta Bhatta. During his period he was renowned as one of the best mathematicians of India. Famous mathematician Al-Biruni translated and mentioned Bateswar's mathematical verses in his books.

Bateswar wrote three books.

(a) Karansar: He wrote the book in 899 when he was 19 years old.

(b) Bateswar Siddhanta: This work was written in 904 when he was 24 years old. This work contains 1326 verses.

(c) Goal: Complete work had not been discovered. A part of this work is available which contains five chapters.

Bateswar Siddhanta contains eight chapters. Many modern formulas of trigonometry and a Sine – Table are there in chapters two and three. He gave value of $\pi = 3927/1250 = 3.1416$.

Two of the trigonometric formulas of Bateswar obtained from his verses are stated below.

1. $R \sin\phi = \sqrt{\{R^2 - (R\cos\phi)^2\}}$
2. $R \cos\phi = \sqrt{\{R^2 - (R\sin\phi)^2\}}$

where ϕ = latitude & R = Radius

Trigonometry formed an integral part of astronomy for our ancient mathematicians. Astronomy became essential for religions for the calculation of accurate time to celebrate religious festivals in accurate time.

From the time of Aryabhata 1 (476 - 550 AD) trigonometry began to resemble its modern form. The Surya Siddhanta (C. AD 400) is a text on astronomy and time keeping, an idea that appears much earlier as the field of Jyotisha (Vedanga) of the Vedic period. Trigonometric concepts are found in this book. References to trigonometric concepts are also found in the Varahamihira's Pancha Siddhantika dated 575 AD and Brahmagupta's Brahmasphuta Siddhanta dated 628 AD. A detailed and systematic study of the subject was made by Bateswar in the Bateswar Siddhanta. The knowledge of Bhaskara II (1114-1185) in trigonometry was unparalleled in his period all over the world.

Infinite expressions of trigonometric function are found in the work of Madhava and Nilkanta based on Bhaskara's work Siddhanta Shiromani. Subsequently the Indian trigonometry transmitted to Arabs who spread the knowledge to Europe after further refinements.

German mathematician and astronomer Regiomontanus (1436-1476) completed his work De Triangulis Omnimodis on trigonometry in 1464. In it he mentioned: You who wish to study great and wonderful things, who wonder about the movement of the stars, must read these theorems about triangles. Knowing these ideas will open the door to all of astronomy and to certain geometric problems. This book was one of the first text books of trigonometry in Europe. He wrote this book based on the works of Arabian mathematicians.

3.12 Aryabhata II (920-1000 AD):

Aryabhata II was a mathematician and astronomer and author of the eminent work Mahasiddhanta. His treatise consists of eighteen chapters. Like other treatise it was also written in the form of verse in Sanskrit. The initial twelve chapters cover the topics : the longitudes of the planets, lunar and solar eclipses, the estimation of eclipses, the lunar crescent, the rising and setting of the planets, association of the planets with each other and with the stars.

The next six chapters of Mahasiddhanta cover the topics : geometry, geography and algebra for astronomical calculations. He gave the rules to solve the indeterminate equation: $by = ax + c$. He deduced the method to calculate the cube root of a number but it was already given by Aryabhata I. Ancient Indian mathematicians tried their best to give the accurate sine tables since they were essential to calculate the planetary positions accurately. Aryabhata II constructed a sine table which was accurate up to five decimal places.

3.13 Bhaskara II (1114-1185):

Bhaskara II was a talented mathematician and astronomer of ancient India. He was born in Bijapur in modern Karnataka. He is also known to us as Bhaskaracharya. His father Mahesvara was a mathematician, astronomer and astrologer who bestowed his knowledge to his affectionate son Bhaskara. Bhaskara's grandfather and descendants held a hereditary post as a court scholar. Bhaskara's knowledge was passed on to his son Loksamudra. Loksamudra's son set up a school in 1207 for the study of Bhaskara's treatise. Ujjain was the leading mathematical center of medieval India. Bhaskara was the head of an astronomical observatory there.

Bhaskara's main work Siddhanta Shiromani (means crown of treatises) was completed in 1150. It is divided into four parts named Lilavati, Bijaganita, Grahaganita and Goladhyaya. These four sections are considered four independent works and deal with arithmetic, algebra, mathematics of planets, and spheres respectively. His noteworthy work is Karana Kautuhala.

It is said that Bhaskara wrote first part of his work Lilavati to teach his daughter, whose name was also Lilavati. After calculating her horoscope Bhaskara realized that if his daughter is not given in marriage in particular time she would be widowed.

He made a water-clock to calculate accurate time in a room and asked his daughter not to enter the room. He filled a pitcher with water and made a very small hole at the bottom. The pitcher was kept at a small height and a cup was kept below it. When the cup would be filled with drops of water the auspicious time would appear. Out of curiosity Lilavati entered the room and began to watch the operation of the system leaning near it. Suddenly a pearl from her nose-ring dropped into the cup. The cup was tilted and water spilt over it. As Lilavati could not be given in marriage in accurate time she became a widow. This fact is pathetic to us.

Lilavati consists of 277 verses and covers the topics of definitions, arithmetical terms, interest computation, arithmetical and geometrical progressions, plane geometry, solid geometry, the shadow of the gnomon (the pin of a dial, whose shadow points to the hour), methods of solving indeterminate equations and combinations. Lilavati has 13 chapters and covers branches of mathematics, arithmetic, algebra, geometry, trigonometry and mensuration.

The second part of his work Bijaganita (means algebra) has 213 verses in twelve chapters. It discusses zero, infinity, positive and negative numbers, surds (includes evaluating surds), Kuttaka (for solving indeterminate equations and Diophantine equations), quadratic equations, solution of indeterminate equations of the second, third and fourth degree etc. His method for the solutions of the problem $ax^2 + bx + c = y$ is very important.

Trigonometry is essential for astronomical calculations. Bhaskara's knowledge in trigonometry was unparalleled in his period all over the world. He used differential calculus and integral calculus in his work Grahaganita (astronomy) and Goladhyaya (mathematics on the sphere). We the people of this planet should regard him as the father of calculus for his preliminary concepts in this field. His inventions were about five hundred years before the birth of Newton and Leibniz.

Bhaskara made a compass like device known as Yasti-Yantra for determining angles with the help of a calibrated scale. He calculated the length of the sidereal year, the time that is required for the Earth to orbit the sun, as 365.2588 days. The modern measurement with sophisticated instruments is 365.2563 days. The difference is of just 3.6 minutes.

3.14 Madhava (1340-1425):

There was a rich mathematical tradition in ancient and medieval Kerala. Many talented mathematicians and astronomers were born there in ancient and medieval times. Madhava, popularly known as Madhavacharya was from the town of Sangamagrama (present name Irinjalakuda) near Thrissur, Kerala. He was one of the greatest mathematician astronomers of the medieval India, who made pioneering contributions to the study of infinite series, calculus, trigonometry, geometry, and algebra. Madhava is considered the founder of the Kerala School of astronomy and mathematics.

Madhava was referred to by later astronomers as Golavid (Master of Spheres). Among his known works are Lagnaprakarana, Aganita, and Aganitapancanga. His Mahajyanayanaprakara and Madhyamanayanaprakara, for which short commentaries are available, contain novel theorems and computational methods evolved by him and used by later mathematicians. Madhava had composed a comprehensive treatise on astronomy and mathematics, which yet remains to be identified and which may be supposed to contain the numerous single and groups of verses enunciating computational procedures, theorems and formulae which are quoted as Madhava's by later mathematicians.

Madhava's contributions show that calculus and analysis had reached remarkable depth and maturity in India centuries before Newton(1642-1727) and Leibniz(1646-1716). Madhava might be regarded as the first mathematician who worked on analysis. Madhava's contributions are mentioned in several later texts by other astronomers and mathematicians.

3.15 Nilakantha Somayaji (1444-1545):

Nilakantha Somayaji, the centenarian astronomer was born in Kerala. He was a versatile scholar and was the author of several works. His Tantrasamgraha (A.D.1500) is a comprehensive treatise on astronomy, in which he mentioned

Madhava's contribution in calculus. Nilakanta was the astronomer who gave the heliocentric model before Copernicus (1473-1543).The most remarkable work of Nilakantha is Grahapariksakrama. There are about 200verses in this work, he sets out the procedures for the observation of the planets, sometimes with instruments, and for their computation using the data obtained from the observations.

Madhava's own books are not available. We learn from commentaries like the Tantrasamgraha that Madhava had initiated mathematical studies into infinite processes dealing with representations of the trigonometric functions, sines and cosines and inverse tangent, as power series.

A detailed commentary on these contributions, is highlighted in the book Yuktibhasa by Jyeshtadeva (C.1501-1610). This document gives proofs of all the propositions and formulas contained therein. This book can, therefore, be considered today as the first text book on calculus.

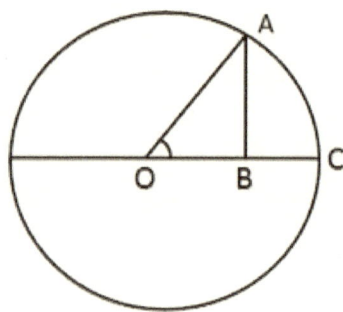

Greek trigonometry (C.150 BCE to 100 CE) was based on relationships obtaining between chords of circles and the angles subtended by them. However, Indian trigonometry, as first seen in Aryabhatiya of Aryabhata, followed up in Pancha Siddhantika by Varahamihira, and in Brahmagupta's Brahmasphuta Siddhanta, is based on half-chords, namely, the sine function, which is also the basis of modern trigonometry. The terms for a triangle is 'trikon' (meaning a three cornered geometric object), for the half-chord AB, 'jya', for the OB, 'kojya', and for segment BC, 'utkramajya' respectively (in figure).

Indian 'jya' becomes 'sine' in Europe, which I have already mentioned. Naturally, the 'kojya' becomes 'cosine' and 'utkramajya' 'versine'.

Besides the many standard formulas of trigonometry, such as $\cos A = \sin(\frac{\pi}{2} - A)$, $\cos 2A = 1 - 2\sin^2 A$, the trigonometric versions of the Pythagorean Theorem, viz,

$\sin^2 A + \cos^2 A = 1$, etc, there are tables giving 24 values of the sine function for angles ranging from $0°$ to $90°$ at equal intervals of $3° 45' = (90/24)°$, which are quite close to the modern values. These tables were used to calculate other values through formulas and, more significantly, by the use of a powerful interpolation formula of Brahamagupta using second – order differences (i.e., differences of the differences of the terms in the given table). These methods of computation were further improved by Madhava who used his approximation formula up to the second power in the power series representation for the sine function to calculate the values accurate up to at least the eighth decimal place.

We get references to Madhava's Mathematical discoveries from Nilakanta Somayaji's 'Tantrasamagraha', in Sanskrit verse, Jyesthadeva's 'Yuktibhasa' in Malayalam and 'Karana Paddhati' by Putuman Somayaji. They attribute to Madhava the discovery of the infinite series expansions for trigonometric functions like sine, cosine and inverse tangent, as well as their finite series approximations up to the second order. Thus rather remarkably, the following infinite series expansions had been discovered by Madhava:

$\sin x = x - x^3/3! + x^5/5! - ... (-1)^{n-1} x^{2n-1}/(2n-1)! + ...;$

$\cos x = 1 - x^2/2! + x^4/4! - ... + (-1)^n x^{2n}/(2n)! + ...;$

$\tan^{-1}(x) = x - x^3/3 + x^5/5 - ... + (-1)^{n-1} x^{2n-1}/(2n-1) + ...;$

Madhava used this expansion for inverse tangent to obtain the first analytical expression for π, now called Madhava-Gregory-Leibniz series, namely,

$\pi/4 = 1 - 1/3 + 1/5 - 1/7 + ... (-1)^{n-1}/(2n-1) + ...$

The Kerala School, or the Nila School after the river Nila in the region, also had knowledge of integration as area under the curve, had an intuitive idea of limits and had given semi-rigorous proofs of the derivatives of certain trigonometric functions. Thus, it appears that mathematical analysis had had its origin in India.

Vasco-da-Gama discovered the sea route to India via the Cape of Good Hope. He arrived at Calicut on May 20, 1498. He opened up a maritime route from Western Europe to India. European missionaries and traders began to arrive in Kerala in greater number. A few of the missionaries and traders were very learned persons who carried the knowledge of calculus to Europe before the birth of Newton (1642-1727) and Leibniz (1646-1716). It is true that Newton and Leibniz made calculus more mature and advanced but Indian mathematicians had given the idea first.

George Gheverghese Joseph of Manchester University and other mathematicians remarked about the Indian discovery of calculus: ".......researches in Britain have gathered evidence that a basic component of calculus developed by mathematicians in Kerala during the 14th century, was passed on to the jesuite scholars who may have carried it to Europe ....and passed on to contemporary mathematicians may have eventually reached Newton".

I have mentioned only a few mathematicians of ancient India but we had about one hundred renowned mathematicians and astronomers in ancient times. Except them thousands of mathematicians and astronomers lived in our country at that time whose names are unknown to us and their works are anonymous.

Many scholars lived in India in Mughal era too. Maharaja Jai Singh II of Jaipur was a regional king under the Mughal Empire. Maharaja Jai Singh was a scholar. He studied books on astronomy and mathematics available at that time in the Persian and Arabic languages. He wrote books on astronomy himself. He was entrusted by Mughal emperor Muhammad Shah the task of revising the calendar and astronomical tables.

In those days there were no sophisticated telescope and other modern instruments for astronomical observation and calculation. So Maha Raja Jai Singh constructed five observatories in Jaipur, Ujjain, Delhi, Mathura, and Varanasi during 1719 to 1734. Calculation by these colossal instruments was not much different from the calculation by modern instruments. Today the observatories are popular tourist attraction and world heritage site declared by UNESCO. These are popularly known as Jantar Mantar.

Teaching system or mode of teaching in our country in ancient times was unique and far better than any system followed in any country even today. A student had to live in the abode of his teacher or guru. There was no special status for son of King or rich persons in those abodes. Students from rich and poor families had to practice of austerities. A guru or teacher used to try their best for the all-round development of his student to bring up him as a perfect man. According to Swami Vivekananda motto of education is:

Education is the manifestation of the perfection already in man.

Our great poet Rabindranath Tagore was also a great educationist. He truly represented the rishis or gurus of ancient India in modern times. In 1901; he founded a school Visva-Bharati at Shantiniketan which was dedicated to emerging Western and Indian philosophy, and education. Visva-Bharati became a university in 1921. It is an educational and cultural center, established in natural surroundings. Visva-Bharati is an ideal institution for the youth. Many best-known personalities like Satyajit Roy, Mrs. Indira Gandhi, Amartya Sen and many other renowned persons were students of this institution. Rabindranath Tagore preached us nationalism along with universalism. His teachings were according to Upanishad, the ideal philosophy of ancient India applicable even today.

Medical sciences: Except mathematics India was advanced in medical sciences from the earliest times. Ayurveda is the earliest school of medicine and health care known to humans. In India Madhusudhan Gupta of Baidyabati, my native town, broke the barrier of superstition or social taboos and dissected a dead-body in Calcutta Medical College on 28 October, 1836 and became the first person in India and Asia to do it from the point of view of modern Western medicine. But in earliest times our doctors (kabirajas) had no superstition, they used to learn about human anatomy by dissecting dead bodies.

India is just like a mini-world. We have places with different climate from very hot Thar Desert to very cold Himalayan range and coastal areas surrounded by seas and very wet Meghalaya. So we have various types of vegetation and medicinal plants and herbs. Except vegetative origin our doctors used to acquire medicine from minerals and from the animal kingdom. Our doctors had about six hundred types of medicines in hand to prescribe according to symptoms of diseases. Use of margosa (neem), turmeric (haldi), sandalwood, tulshi etc. is a tradition in Indian household.

Most of our doctors were monks and treatment of patient was a dedication to humankind for them. "Not for self, not for fulfillment of earthly desires of gain, but solely for the good of suffering humanity should you treat your patients, and so excel all". This was the first code of conduct for doctors. Even to this day, almost the same code of conduct is followed in the medical profession world-wide.

Dhanwantari was an excellent doctor of gods, he excelled both in medicine and surgery. Even to this day excellent doctors are called "Dhanwantari" by his patients.

There was an unparalleled doctor in India in modern times who was the chief minister of West Bengal from 1948 to 1962 till his death. Now and again, a genius like Dr. Bidhan Chandra Roy is born. He was regarded as a "Dhanwantari" in India. He went to England with rupees one thousand and two hundred only as his financial condition was not good. Such was his tenacity and talent that he achieved the highest honor in medicine and surgery in only two years. It was only possible for Dr. Roy to become MRCP and FRCS, within a short period of two years.

After returning from England Dr. Roy started practice in Calcutta (now Kolkata). He began to get patients even from neighboring countries. Many contemporary leaders of India and the members of their families were his patients. Desbandhu Chittaranjan Das, Matilal Nehru, Mahatma Gandhi, Ballavbhi Patel, Moulana Abul Kalam Ajad, Jawaharlal Nehru and his daughter Indira Gandhi and many other renowned persons were his patients.

Once Mahatma Gandhi was confined in Aga Khan Palace in Pune. He even stopped taking any medicine so his health was deteriorating rapidly. Dr. Roy reached there for his treatment.

Gandhi was adamant, he refused Dr. Roy's request. Then Dr. Roy said, "I have not come here to treat Mohandas Karamchand Gandhi, I am here to treat a person who is the leader of my country's forty crore people." Gandhi said, "You argue like a lawyer, all right give me medicine, I shall take". Such incident happened several times during Gandhi's fast and he could not refuse Dr. Roy's request.

Among Dr. Roy's admirers were John. F. Kennedy, president of America and Clement Attlee, prime minister of Britain.

I was a child when he was the chief minister of West Bengal. I have heard from my grandfather that Dr. Roy could diagnose a patient from his body odour or odour in a patient's room. He used to treat sixteen patients in the morning without taking any fees though he had the great responsibility of chief ministership. Dr.Roy remained bachelor and dedicated his life for the service of humankind. He donated his palatial residence for a hospital.

Robert Koch had discovered the cholera bacteria in 1884 but Dr. Sambhu Nath De unearthed the fact in 1959 that it was a toxin produced by the bacteria that triggered the loss of fluid from the body and caused thickening of blood, eventually leading to death.

The oral rehydration therapy which has saved millions of cholera patients over the years is a direct outcome of his discovery. He received international recognition for this work and nominated by the then Nobel laureate Joshua Lederberg for the Nobel Prize.

Our ancient system of medicine Ayurveda has been traced back to 5000 BC. Ayurveda means 'complete knowledge for long life.' Our ancient doctor Charaka compiled all earlier works in the field of medicine and surgery in his work 'Charaka Samhita' written in first century AD or earlier. He described cause of a disease, symptoms of a disease and particular medicine of that disease.

Sushruta was a professor of medicine at Varanasi. Sushruta's medical treatise 'Sushruta Samhita' consists of 184 chapters, 1120 conditions are listed, including injuries and illness relating to ageing and mental illness. This work has description of procedures on various forms of surgery including rhinoplasty, the repair of torn ear lobes, perineal lithotomy, cataract surgery etc. The Sushruta Samhita described 125 surgical instruments, 300 surgical procedures and classifies human surgery in eight categories.

Unani medicine is much close to Ayurveda. It progressed during sultanate and Mughal periods. Ancient Egyptians, Babylonians and people of China were also very advanced in the field of medicine and surgery.

Yoga: Ayurveda gave us medicine and treatment to get rid of diseases. Our forefathers gave us another exclusive system to prevent diseases, i.e. prevention is better than cure. Yoga is an ancient mind body practice which originated in India. Aerobic exercises are good for health but yoga is more effective than Aerobic exercises. Yoga is a system which prevents as well as cures diseases. Yoga incorporates physical, mental, and spiritual elements and has been shown in several studies to be effective in improving cardiovascular risk factors, with reduction in the risk of heart attacks and strokes. Yoga plays a vital role in reducing stress which leads to positive impacts on neuroendocrine status, metabolic and cardio-vagal function.

The most classical text of yoga, the Patanjali yoga sutra, is composed of 196 verses. According to Patanjali, asana was and is meant to use the body as a tool to train the mind and the senses. Patanjali yoga sutra was translated in Arabic by the renowned scholar Al-Beruni, who named it "Kitab Patanjal". Thus yoga crossed the barrier of religion like mathematics. Ancient scholars of Iran and Iraq region translated Indian mathematics and other works in Arabic. It's interesting to note that the works are called books of original writers of Indian origin and no credit is robbed from original writers of Indian originality. Many of the original works have been lost from India. We can know about them from Arabic translations. We owe a lot to the Arabian scholars.

Music: We had have classical music from times immemorial. According to doctors' opinion Indian classical music has healing effect on patients.

Metallurgy: Our metallurgists had advanced knowledge and practical skills in extraction of iron, zinc, copper and refining of gold. We have got a dancing girl figure cast in bronze of 800gm and 10.5cm in height from the excavations in the Indus Valley. This figure is the pride of Indians as it had been cast about 5000 years ago. The iron and steel industry throughout India was a tradition from 1300 BC. Damascus was then famous for making the best quality swords. They imported steel from India to manufacture these swords. Asoka - the Great erected pillars on which were inscribed his understanding of religious doctrines. These iron pillars are in the open air for more than two thousand years. The pillars are as it is as they were made of rust-proof steel. Is it not an enigma?

Legend has it that one of the gifts that Alexander the great took from king Porus was a ball of steel weighing about 15 kilograms.

I heard from Professor Ananta Lal Thakur of Patna University who lived at Baidyabati that Alexander had taken many books from Indian libraries during his raid. Alexander was taught by Aristotle and it is possible for a student of Aristotle to take books from India but I have not read this fact from any book.

Our Bengal was also very advanced in metallurgy. There is a

big cannon named **Dalmadal** in the open air at Bishnupur in the district of Bankura. It was used to prevent Bargi raid in 1742 during the reign of king Gopal Singha. It is 3.8 meter long and weighing 296 maunds (1 maund = 37.3242 kg.). Speciality of this cannon is rust proof.

I could stop writing and finish my book in previous paragraph but it would remain incomplete if I do not mention two mathematicians of Bengal.

3.16 Radhanath Sikdar (1813-1870):

Radhanath Sikdar was a talented student from boyhood. Like Ramanujan he had also passed through poverty and starvation during student life. He got admission in Hindu College, Calcutta (Kolkata) in 1824. Henry Louis Vivian Derozio was his English teacher in the college. Influence of Derozio moulded Radhanath's progress. Dr. Jan Titalor, mathematics teacher of his college, first noticed his talent in mathematics. Dr. Jan presented Radhanath Principia of Newton for thorough study.

Radhanath joined as a Surveyor in the Great Trigonometrical Survey of India on 20th December, 1831. George Everest was then the Surveyor General of India. In 1852, Radhanath Sikdar calculated the height of peak-XV of the Himalayas as 29002.3 feet and became sure that it was the highest peak in the world. He christened the peak Mount Everest in honor of his ex-boss George Everest.

3.17 Ashutosh Mukherjee (1864-1925) :

Unlike other children Ashutosh Mukherjee was interested in mathematics from childhood. In later life he became a keen mathematician.

He was Vice-Chancellor of Calcutta University (1906-1914, 1921-1923) and during his tenure he made the university as an ideal educational institution in the country. During his time renowned scientists Dr. Prafulla Chandra Roy, Dr. Jagodish Chandra Bose were professors at the university.

When Ashutosh Mukherjee was a student of B.A 1st year, a mathematics research paper of him was published in Messenger of Mathematics, a research journal published in Britain. After that several research papers of his were published in different mathematics journals.

Fields Medal: The Fields Medal, named after the renowned Canadian mathematician Professor John Charles Fields is being regularly awarded every four years during the international congress of the International Mathematical Union (IMU). It was conceived at the 1924 congress in Toronto, in which he was the secretary. He also donated funds for it.

Those who received the prestigious Fields Medal in 2014 include Manjul Bhargava, a Canadian-American of Indian Origin; Artur Avila, a Brazilian French; Martin Hairer, an Austrian and Maryam Mirzakhani, an Iranian.

Manjul Bhargava maintains the rich mathematical tradition of India. Many famous mathematicians were born in Iran in ancient times. Maryam Mirzakhani, the first female Fields medalist, maintains the rich heritage of her country Iran. We are proud of both Manjul and Maryam.

The recipients of the Fields Medal for 2018 have been published. New Delhi-born Indian-Australian mathematician Akshay Venkatesh, who is currently teaching at Stanford University is one of them. The other Fields medalists this year are Peter Scholze, of the University of Bonn; Caucher Birkar, of the University of Cambridge in England; and Alessio Figalli, of the Swiss Federal institute of Technology in Zurich.

Questions and Answers in quiz game

1. When and where was Ramanujan born in India?
2. Who were Ramanujan's Parents?
3. Where was Ramanujan's ancestral home?
4. From which Primary School did Ramanujan stand first in the district in 1897 at the age of ten?
5. Who was the headmaster of the Town High School in Kumbakonam who first observed Ramanujan's prodigious mathematical ability?
6. Name the subject he failed for which his scholarship was stopped.
7. In which year with whom Ramanujan was married?
8. In which office did he join as a grade IV clerk in March 1912?
9. To which fellow mathematician did Professor Hardy show Ramanujan's first set of theorems, received in January 1913?
10. From which Goddess did he get consent in dream before going to England?
11. What was the name of the ship on which he sailed to England on March 17, 1914?
12. What was the subject of Ramanujan that was first

Ancient Indian Mathematicians

published in London Mathematical Society in 1914?

13. Who came from England and persuaded Ramanujan for going to Cambridge University?

14. Who was the tutor of Ramanujan in Cambridge University?

15. What is Ramanujan Number?

16. Who was the chairman of the Royal Society when Ramanujan was elected F.R.S. in 1918?

17. Which eminent Indian statistician and later member of the planning commission was a bosom friend of Ramanujan during his stay at Cambridge?

18. What name did Ramanujan give to the family of function studied by him in his last days in Madras?

19. Who was the first Indian F.R.S. in 1841?

20. Who was the first Indian F.R.S. in mathematics in 1918?

21. When Ramanujan breathed his last?

22. Who properly used Zero in mathematics?

23. Which Indian mathematician introduced infinity (∞)?

24. In which country decimal system was introduced for counting?

25. Which country's counting system is in vogue all over the modern world?

Answers of the Quiz

1. Ramanujan was born on December 22, 1887 at Erode
2. His father and mother were Srinivasa Iyengar and Komalatammal.
3. Kumbakonam.
4. Kangayan Primary School.
5. S.Krishnaswami Iyer.
6. English Composition.
7. Ramanujan was married to 9-year old Janaki, on 14 July 1909.
8. Madras Port Trust.
9. John Edensor Littlewood.
10. Namagiri.
11. S.S. Nevasa.
12. Modular Equations and Approximations to Pi.
13. Eric Harold Neville.
14. E.W. Barnes.
15. 1729
16. J.J. Thomson, Nobel Laureate in Physics.
17. Prasantha Chandra Mahalanobis.
18. Mock Theta functions.
19. Adraseer Cursetjee was the first Indian fellow of the Royal Society on 27 May 1841.
20. Srinivasa Ramanujan Iyengar.

21. Ramanujan breathed his last on 26 April 1920.
22. Aryabhata 1
23. Bhaskara II
24. India.
25. Indian Counting system is in vogue all over the world.

References

1. Science Reporters
2. The Man who knew infinity -Robert Kanigel
3. Ignited Minds-APJ Abdul Kalam
4. Cosmos- Carl Sagan
5. A Brief History of Time- Stephen W. Hawking
6. The Cosmic Detective- Dr. Mani Bhaumik
7. Einstein for Everyone- Robert L. Piccioni
8. The story of my experiments with truth-M.K.Gandhi
9. Mathability- Shakuntala Devi
10. A dictionary of Sanskrit -English Technical Terms - Volume 1, mathematics - Compiled & Edited by - Prof. Pradip Kumar Majumdar
11. India Through The Ages- Bharat Bhushan Gupta
12. 1001 Inventions that changed the world -General Editor Jack Challoner, Preface by Trevor Baylis
13. Prachin Bharate Ganitcharcha- Dr. Pradip Kumar Majumdar (Bengali Book)
14. Bismrita Bharat- Shree Rajib Chakraborty (Bengali Book)
15. Ramanujan O Prachin Bharate Ganitcharcha- Swapan Bandyopadhyay (Bengali Book)
16. Master Mind India – Siddhartha Basu
17. Articles published in THE TIMES OF INDIA (NEWS PAPER – KOLKATA)

18. Various Articles written by Scholars collected from Internet.
19. Mathematical Hand book – M.Vygodsky
20. RAMANUJAN, Twelve lectures on subjects suggested by his life and work – G.H. Hardy.
21. Concepts of Physics – H C Verma.
22. What is Mathematics? – Balkrishna Shetty.
23. The Book of Numbers – Shakuntala Devi.
24. The Aryabhatiya of Aryabhata – Walter Eugene Clark
25. Studies of Trigonometry in Ancient India. – Dr.Pradip Kumar Majumdar.
26. Articles published in THE ANANDA BAZAR PATRIKA written by Pathik Guha (NEWS PAPER - KOLKATA)
27. The Crest of the Peacock - George Gheverghese Joseph
28. The Discovery of India - Jawaharlal Nehru.

Appendix

Geometric figures for the Construction of altars

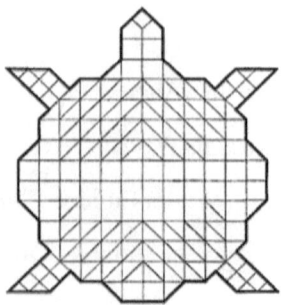

Figure 10: The kūrma citi: layers 1, 3, and 5.

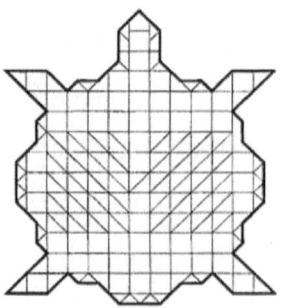

Figure 11: The kūrma citi: layers 2 and 4.

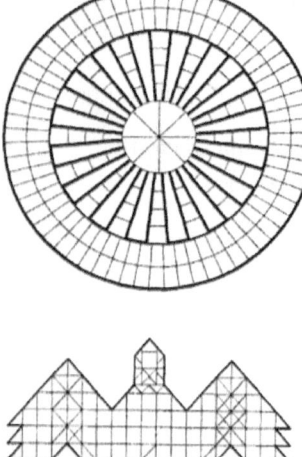

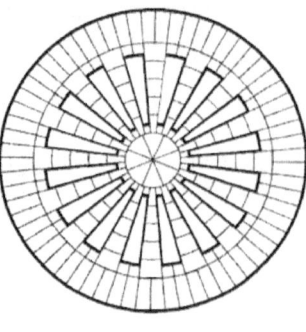

Figure 7: The śyena citi: layers 1, 3, and 5.

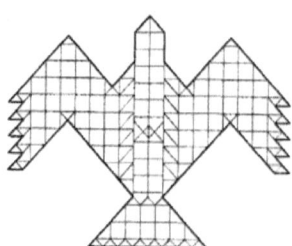

Figure 8: The śyena citi: layers 2 and 4.

Ancient Indian Mathematicians

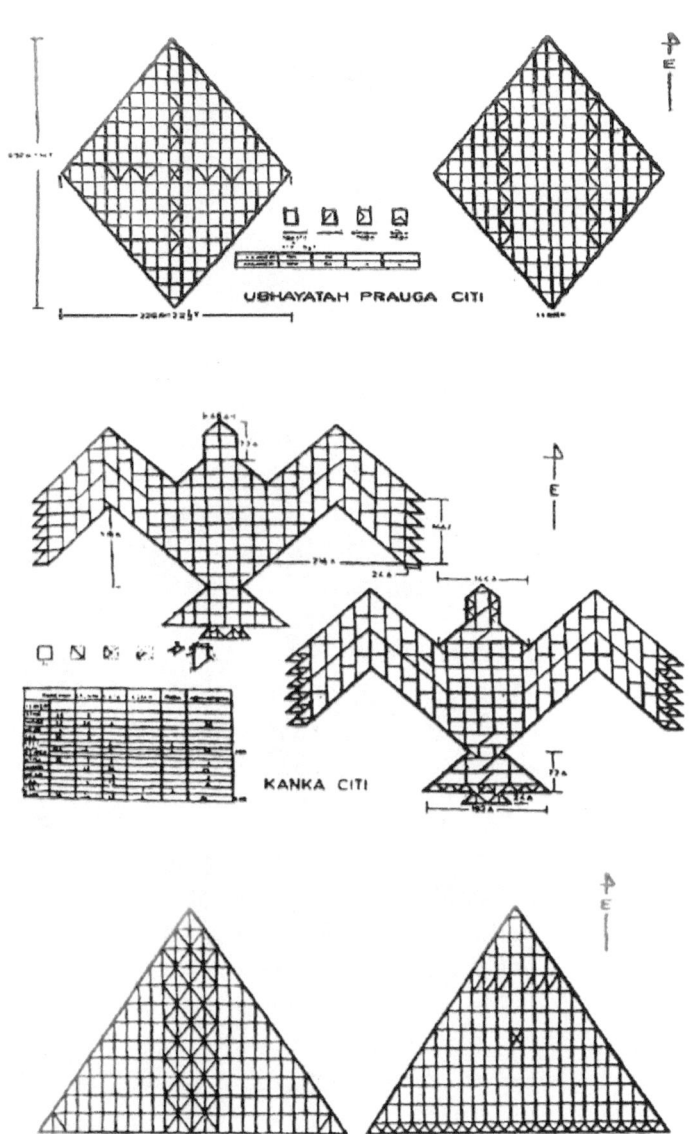

Swapan Banerjee's passion for reading and writing and love for mathematics and science subjects can be traced back to the days of his school in Beliatore High School, in the district of Bankura, West Bengal, India.

The author, deeply interested in Indian History and her Heritage, has made it a lifelong passion to establish Indian contribution to the global field of Mathematics, Science, Astronomy and Philosophy.

A graduate in pure science from the Missionary College at Serampore and with a technical degree in teaching method, he started his career as a teacher in a senior school in the beautiful North Island of the Andaman. Later he joined the financial sector of a government owned bank. Now he is continuing his search for knowledge in oriental studies.

The author is deeply involved in the studies of natural science and mysteries of the universe.